D1220828

Volume 69

Tissue-Specific Vascular Endothelial Signals and Vector Targeting, Part B

Advances in Genetics, Volume 69

Serial Editors

Theodore Friedmann
University of California at San Diego, School of Medicine, USA
Jay C. Dunlap
Dartmouth Medical School, Hanover, NH, USA
Stephen F. Goodwin
University of Oxford, Oxford, UK

Volume 69

Tissue-Specific Vascular Endothelial Signals and Vector Targeting, Part B

Edited by

Renata Pasqualini

Departments of Experimental Diagnostic Imaging,
Genitourinary Medical Oncology, Cancer Biology and
David H. Koch Center
The University of Texas M.D. Anderson Cancer Center,
Houston, Texas, USA

AMSTERDAM • BOSTON • HEIDELBERG • LONDON
NEW YORK • OXFORD • PARIS • SAN DIEGO
SAN FRANCISCO • SINGAPORE • SYDNEY • TOKYO
Academic Press is an imprint of Elsevier

ELSEVIER

LIBRARY
FRANKLIN PIERCE UNIVERSITY
RINDGE, NH 03461

Academic Press is an imprint of Elsevier

525 B Street, Suite 1900, San Diego, CA 92101-4495, USA
30 Corporate Drive, Suite 400, Burlington, MA 01803, USA
32 Jamestown Road, London, NW1 7BY, UK
Radarweg 29, POBox 211, 1000 AE Amsterdam, The Netherlands

First edition 2010

Copyright © 2010 Elsevier Inc. All rights reserved.

No part of this publication may be reproduced, stored in a retrieval system
or transmitted in any form or by any means electronic, mechanical, photocopying,
recording or otherwise without the prior written permission of the publisher

Permissions may be sought directly from Elsevier's Science & Technology Rights
Department in Oxford, UK: phone (+44) (0) 1865 843830; fax (+44) (0) 1865 853333;
email: permissions@elsevier.com. Alternatively you can submit your request online by
visiting the Elsevier web site at http://www.elsevier.com/locate/permissions, and selecting
Obtaining permission to use Elsevier material.

Notice

No responsibility is assumed by the publisher for any injury and/or damage to persons or
property as a matter of products liability, negligence or otherwise, or from any use or operation
of any methods, products, instructions or ideas contained in the material herein. Because of
rapid advances in the medical sciences, in particular, independent verification of diagnoses and
drug dosages should be made.

ISBN: 978-0-12-375022-8
ISSN: 0065-2660

For information on all Academic Press publications
visit our website at elsevierdirect.com

Printed and bound in USA

10 11 12   10 9 8 7 6 5 4 3 2 1

Working together to grow
libraries in developing countries

www.elsevier.com | www.bookaid.org | www.sabre.org

ELSEVIER   BOOK AID
             International   Sabre Foundation

Contents

6   On the Synergistic Effects of Ligand-Mediated and Phage-Intrinsic Properties During *In Vivo* Selection   115

Wouter H. P. Driessen, Lawrence F. Bronk, Julianna K. Edwards, Bettina Proneth, Glauco R. Souza, Paolo Decuzzi, Renata Pasqualini, and Wadih Arap

7   Strategies for Targeting Tumors and Tumor Vasculature for Cancer Therapy   135

Prashanth Sreeramoju and Steven K. Libutti

Contributors

Numbers in parentheses indicate the pages on which the authors' contributions begin.

Wadih Arap (31, 97, 115) David H. Koch Center, The University of Texas M.D. Anderson Cancer Center, Houston, Texas, USA, and Department of Experimental Diagnostic Imaging, The University of Texas M.D. Anderson Cancer Center, Houston, Texas, USA, and Department of Genitourinary Medical Oncology, The University of Texas M.D. Anderson Cancer Center, Houston, Texas, USA; Department of Cancer Biology, The University of Texas M.D. Anderson Cancer Center, Houston, Texas, USA

Zaver M. Bhujwalla (1) JHU ICMIC Program, Russell H. Morgan Department of Radiology and Radiological Science, The Johns Hopkins University School of Medicine, Baltimore, Maryland, USA, and The Sidney Kimmel Comprehensive Cancer Center, The Johns Hopkins University School of Medicine, Baltimore, Maryland, USA

Lawrence F. Bronk (115) David H. Koch Center, The University of Texas M.D. Anderson Cancer Center, Houston, Texas, USA

Paolo Decuzzi (31, 115) Department of Nanomedicine and Biomedical Engineering, The University of Texas Health Science Center Houston, Houston, Texas, USA, and Center of Bio-/Nanotechnology and Engineering for Medicine, University of Magna Graecia, Catanzaro, Italy

Wouter H. P. Driessen (31, 115) David H. Koch Center, The University of Texas M.D. Anderson Cancer Center, Houston, Texas, USA, and Department of Genitourinary Medical Oncology, The University of Texas M.D. Anderson Cancer Center, Houston, Texas, USA; Department of Cancer Biology, The University of Texas M.D. Anderson Cancer Center, Houston, Texas, USA

Julianna K. Edwards (115) David H. Koch Center, The University of Texas M.D. Anderson Cancer Center, Houston, Texas, USA

Mauro Ferrari (31) Department of Nanomedicine and Biomedical Engineering, The University of Texas Health Science Center, Houston, Texas, USA, and Department of Experimental Therapeutic, The University of Texas M.D. Anderson Cancer Center, Houston, Texas, USA;

Department of Biomedical Engineering, Rice University, Houston, Texas, USA

Biana Godin (31) Department of Nanomedicine and Biomedical Engineering, The University of Texas Health Science Center, Houston, Texas, USA

Amin Hajitou (65) Department of Gene Therapy, Section/Division of Infectious Diseases, Faculty of Medicine, Imperial College London, Wright-Fleming Institute, St Mary's Campus, Norfolk Place, London, United Kingdom

Sei-Young Lee (31) Department of Nanomedicine and Biomedical Engineering, The University of Texas Health Science Center, Houston, Texas, USA, and Department of Mechanical Engineering, The University of Texas at Austin, Austin, Texas, USA

Steven K. Libutti (135) Montefiore Medical Center, Albert Einstein College of Medicine, Bronx, New York, USA

Xiaoli Lu (83) Department of Microbiology and Molecular Genetics, 427 Bridgeside Point II, 450 Technology Drive, University of Pittsburgh School of Medicine, Pittsburgh, Pennsylvania, USA

Zhibao Mi (83) Department of Microbiology and Molecular Genetics, 427 Bridgeside Point II, 450 Technology Drive, University of Pittsburgh School of Medicine, Pittsburgh, Pennsylvania, USA

Renata Pasqualini (31, 97, 115) David H. Koch Center, The University of Texas M.D. Anderson Cancer Center, Houston, Texas, USA, and Department of Experimental Diagnostic Imaging, The University of Texas M.D. Anderson Cancer Center, Houston, Texas, USA, and Department of Genitourinary Medical Oncology, The University of Texas M.D. Anderson Cancer Center, Houston, Texas, USA; Department of Cancer Biology, The University of Texas M.D. Anderson Cancer Center, Houston, Texas, USA

Arvind P. Pathak (1) JHU ICMIC Program, Russell H. Morgan Department of Radiology and Radiological Science, The Johns Hopkins University School of Medicine, Baltimore, Maryland, USA, and The Sidney Kimmel Comprehensive Cancer Center, The Johns Hopkins University School of Medicine, Baltimore, Maryland, USA

Marie-France Penet (1) JHU ICMIC Program, Russell H. Morgan Department of Radiology and Radiological Science, The Johns Hopkins University School of Medicine, Baltimore, Maryland, USA

Bettina Proneth (31, 115) David H. Koch Center, The University of Texas M.D. Anderson Cancer Center, Houston, Texas, USA, and Department of Genitourinary Medical Oncology, The University of Texas M.D. Anderson Cancer Center, Houston, Texas, USA; Department of Cancer Biology, The University of Texas M.D. Anderson Cancer Center, Houston, Texas, USA

Paul D. Robbins (83) Department of Microbiology and Molecular Genetics, 427 Bridgeside Point II, 450 Technology Drive, University of Pittsburgh School of Medicine, Pittsburgh, Pennsylvania, USA

Rolando Rumbaut (31) Children's Nutrition Research Center, Baylor College of Medicine, Houston, Texas, USA

Masanori Sato (97) David H. Koch Center, The University of Texas M.D. Anderson Cancer Center, Houston, Texas, USA

Glauco R. Souza (115) David H. Koch Center, The University of Texas M.D. Anderson Cancer Center, Houston, Texas, USA

Prashanth Sreeramoju (135) Montefiore Medical Center, Albert Einstein College of Medicine, Bronx, New York, USA

Srimeenakshi Srinivasan (31) Department of Nanomedicine and Biomedical Engineering, The University of Texas Health Science Center, Houston, Texas, USA

Virginia J. Yao (97) David H. Koch Center, The University of Texas M.D. Anderson Cancer Center, Houston, Texas, USA

Maliha Zahid (83) Department of Microbiology and Molecular Genetics, 427 Bridgeside Point II, 450 Technology Drive, University of Pittsburgh School of Medicine, Pittsburgh, Pennsylvania, USA

1

MR Molecular Imaging of Tumor Vasculature and Vascular Targets

Arvind P. Pathak,[*,†] **Marie-France Penet,**[*] **and Zaver M. Bhujwalla**[*,†]

[*]JHU ICMIC Program, Russell H. Morgan Department of Radiology and Radiological Science, The Johns Hopkins University School of Medicine, Baltimore, Maryland, USA
[†]The Sidney Kimmel Comprehensive Cancer Center, The Johns Hopkins University School of Medicine, Baltimore, Maryland, USA

0065-2660/10 $35.00
DOI: 10.1016/S0065-2660(10)69010-4

ABSTRACT

Tumor angiogenesis and the ability of cancer cells to induce neovasculature continue to be a fascinating area of research. As the delivery network that provides substrates and nutrients, as well as chemotherapeutic agents to cancer cells, but allows cancer cells to disseminate, the tumor vasculature is richly primed with targets and mechanisms that can be exploited for cancer cure or control. The spatial and temporal heterogeneity of tumor vasculature, and the heterogeneity of response to targeting, make noninvasive imaging essential for understanding the mechanisms of tumor angiogenesis, tracking vascular targeting, and detecting the efficacy of antiangiogenic therapies. With its noninvasive characteristics, exquisite spatial resolution and range of applications, magnetic resonance imaging (MRI) techniques have provided a wealth of functional and molecular information on tumor vasculature in applications spanning from "bench to bedside". The integration of molecular biology and chemistry to design novel imaging probes ensures the continued evolution of the molecular capabilities of MRI. In this review, we have focused on developments in the characterization of tumor vasculature with functional and molecular MRI. © 2010, Elsevier Inc.

I. INTRODUCTION

The two major mechanisms resulting in the formation of blood vessels are vasculogenesis and angiogenesis. Vasculogenesis describes the establishment of the vasculature during embryogenesis and development. Angiogenesis, a term coined by Dr. John Hunter in 1794 to describe the formation of new blood vessels from extant vasculature (Hunter, 1794), is a process that occurs in both the embryo and the adult (Carmeliet, 2005). In 1865, Rudolf Virchow made the observation that tumors have distinct capillary networks (Virchow, 1863), which was followed by the first systematic studies of the tumor vasculature by Goldman in 1907 (Goldman, 1907) and by Lewis in 1927 who determined that the tumor environment has a profound effect on the architecture of angiogenic vessels (Lewis, 1927). (See Ribatti's (2009) comprehensive treatise on the history of tumor angiogenesis research.) A century later, the observation that adult angiogenesis was a hallmark of pathologies ranging from cancer to diabetic retinopathy (Jain and Carmeliet, 2001) led Folkman (1971) to posit in his seminal paper that solid tumor growth was "angiogenesis-dependent". In it, he also introduced the concept of "antiangiogenic" therapy or the idea that solid tumor growth could be arrested by preventing the recruitment and formation of de novo blood vessels. In the decades since, a comprehensive understanding of the molecular mechanisms regulating tumor angiogenesis has emerged, resulting in the identi-fication of a slew of angiogenesis inhibitors, many of which are currently in clinical trials (Folkman, 2007).

In a careful study of carcinoma of the bronchus, Thomlinson and Gray observed that the onset of necrosis occurred at approximately 160 μm from the nearest vessel, a distance calculated to be the diffusion limit of oxygen. Based on these data, they predicted the presence of hypoxia in tumors that would lead to radioresistance (Thomlinson and Gray, 1955). Almost four decades later, with the discovery of the hypoxia inducible factor-1 (HIF-1), and its role as a transcrip-tional regulator of an ever-increasing list of genes (Semenza, 2010), tumor hypoxia resulting from the chaotic tumor vasculature has been implicated in metabolism, angiogenesis, invasion, metastasis, and drug resistance (Bertout et al., 2008).

Since Clark et al. (1931) created some of the earliest images of neovas-cularization in transparent rabbit ear chambers in the 1930s, advances in physics (e.g., new imaging methods), chemistry (e.g., the synthesis of novel imaging probes), and biology (e.g., development of innovative gene reporter systems and the identification of novel targets) have ushered in a new era in the characteri-zation of angiogenesis and antiangiogenic therapy using imaging (McDonald and Choyke, 2003; Pathak et al., 2008a). Even a century ago, the importance of "individualizing cancer treatment" and "penetrating into the darkness of physio-logical conditions existing in tumor growths" was recognized in the prescient remarks made by E. Goldman in 1907 (Goldman, 1907).

The importance of tumor vasculature in several phenotypic characteristics of cancer, as well as in drug delivery and metastasis, has become very evident, and angiogenic or vascular targeting is meeting with some success as a potential treat-ment for cancer (Folkman, 2007; Neri and Bicknell, 2005). These developments have not only necessitated the genomic and functional characterization of individ-ual tumors to identify specific molecular targets, but also the ability to noninvasively detect the spatial and temporal response to these new targeted therapies. Noninva-sive imaging techniques, the availability of "smart probes" as well as molecular strategies such as the use of small interfering RNA to downregulate specific targets are playing an increasingly important role in this era of targeted molecular medicine. The purpose of this review is to describe recent advances in MRI as applied to tumor vasculature characterization and targeting.

II. STRUCTURAL, FUNCTIONAL, AND MOLECULAR CHARACTERISTICS OF TUMOR VASCULATURE

Studies of tumor vascular morphology have identified a variety of structural and functional differences between tumor and normal vasculature (see Konerding et al., 2000 for a comprehensive review). Tumor-induced blood vessels are typically sinusoidal, exhibit discontinuous basement membranes, and lack tight endothelial cell junctions making them highly permeable to macromolecules. Other characteristics of the tumor vasculature include (i) spatial heterogeneity and loss of branching hierarchy, (ii) arteriovenous shunts, (iii) acutely and transiently collapsing vessels, (iv) poor differentiation and a lack of smooth muscle cell lining, and (v) an inability to match the elevated metabolic demand of cancer cells, resulting in areas of hypoxia and necrosis.

Pioneering work by Jain, Vaupel, and others has demonstrated that structural anomalies of the tumor vasculature result in altered hemodynamics, blood rheology, and tumor blood flow (Jain, 1988; Vaupel *et al.*, 1989). Figure 1.1 summarizes the bidirectional relationships between the anomalous aspects of the tumor vasculature and the resulting pathophysiological and molecular perturbations in the tumor microenvironment.

In addition to angiogenesis-dependent pathways, nonangiogenic pathways for tumor growth have also been observed. Of these, vascular cooption and vasculogenic mimicry are the most well known. In a landmark study, Holash

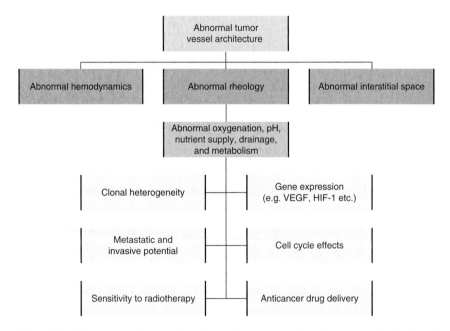

Figure 1.1. Schematic to illustrate how abnormal tumor vessel architecture results in altered hemodynamics (blood flow), blood rheology, and elevated interstitial fluid pressure. These alterations in turn profoundly affects the tumor microenvironment (i.e., oxygenation, pH, etc.), which in turn modulates multiple phenomena ranging from gene expression to sensitivity to radiotherapy. Adapted from Molls and Vaupel (2000).

et al., (1999) demonstrated that in contrast to the prevailing view that most tumors begin as avascular masses, a subset of tumors initially grew by "coopting" existing host blood vessels. This coopted host vasculature did not immediately undergo angiogenesis but initially regressed, leading to an avascular tumor with massive tumor cell loss. Eventually, the remaining tumor was rescued by robust angiogenesis at the tumor rim. Maniotis *et al.* (1999) described another mode of vascular channel formation which was dubbed "vasculogenic mimicry" to highlight the fact that parts of the microcirculation of aggressive uveal melanomas consist of channels lined by a layer of extracellular matrix and the tumor cells themselves. Chang *et al.* (2000) have described the formation of "mosaic vessels" in a colon carcinoma model, wherein both tumor and endothelial cells contributed to vascular tube formation.

The heterogeneity of tumor vasculature can pose a formidable clinical challenge. The structural and functional deficiencies of the tumor vasculature profoundly impact drug delivery, radiosensitivity, proliferation rate, invasion, metastases, and the metabolic micromilieu (pO2, pH, energy status; Konerding *et al.*, 2000) of the tumor. However, as discussed subsequently, this heterogeneity of the tumor vasculature also presents an opportunity for the identification of novel drug targets that can be exploited to develop tumor vasculature-selective therapeutic strategies (Neri and Bicknell, 2005).

As summarized in several excellent reviews (Baluk *et al.*, 2005; McDonald and Choyke, 2003; Munn, 2003), the physiology and organization of the tumor vasculature at the spatial scale of the endothelial cell is also abnormal. Such abnormalities include gaps in interendothelial tight junctions resulting in loss of barrier function, loose or no association with pericytes, and a compromised or incomplete basement membrane (Kalluri, 2003). This microscopic picture is further complicated by the phenomena of vasculogenic mimicry and mosaic vessels described earlier. However, at this scale, the diversity of surface proteins selectively expressed by angiogenic tumor vessels has been exploited as targets for novel contrast agents using a range of imaging modalities (McDonald and Choyke, 2003). Some of these epitopes on angiogenic endothelial cells and basement membrane components have also been tested for tumor vessel selective drug-targeting strategies as summarized in Table 1.1 (Molema, 2005). The review by Langenkamp and Molema (2009) lists endothelial cell-specific genes used to identify tissue microvasculature, while a landmark paper by Croix *et al.* (2000) demonstrates the diversity of the genes expressed in human tumor endothelium. Both imaging receptor expression and vascular targeting are discussed in detail in ensuing sections.

Conventional image contrast in radiologic images provides differences in the level of brightness or intensity of parts of the image corresponding to anatomically or physiologically different locations. While traditionally such contrast has aided in the differentiation of normal from pathologic tissue, such

Table 1.1. Epitopes on Angiogenic Endothelial Cells and Basement Membrane Components that may serve as Targets for Tumor Vasculature-Selective Drug-Targeting Strategies

Target	References
30.5 kDa antigen	Hagemeier *et al.* (1986)
CD34	Schlingemann *et al.* (1990)
VEGF–VEGFR complex[a]	Brown *et al.* (1993)
Endosialin	Rettig *et al.* (1992)
Selectins[a]	Nguyen *et al.* (1993)
αv integrins[a]	Brooks *et al.* (1994)
Endoglin[a]	Burrows *et al.* (1995)
Tie-2	Sato *et al.* (1995)
Angiostatin receptor	Moser *et al.* (1999)
MMP-2/MMP-9[a]	Koivunen *et al.* (1999)
CD13/Aminopeptidase N[a]	Pasqualini *et al.* (2000)
Endostatin receptor	Karumanchi *et al.* (2001)
TEM 1/5/8	Croix *et al.* (2000)
VE cadherin cryptic epitope[a]	Corada *et al.* (2002)
CD44v3	Forster-Horvath *et al.* (2004)
Annexin A1	Oh *et al.* (2004)
Inducible target	
P-selectin	Hallahan *et al.* (1998)
Extracellular matrix target	
EDB-Fn[a]	Tarli *et al.* (1999)
Basement membrane component	Epstein *et al.* (1995)

Adapted with permission from Molema (2005).

[a]Denotes the target molecules experimentally employed for tumor vascular drug-targeting strategies. EDB-Fn, EDB-oncofetal domain of fibronectin; MMP, matrix metalloproteinase; TEM, tumor endothelial marker; VEGF(R), vascular endothelial growth factor (receptor).

images provide little information regarding the functional or molecular status of a lesion. For example, one cannot infer the angiogenic status of the tumor vasculature from conventional imaging (Pathak *et al.*, 2008a). A wide array of noninvasive imaging modalities has been used to image the tumor vasculature. These include X-ray computed tomography (CT), MRI, positron emission tomography (PET), single-photon emission computed tomography (SPECT), ultrasound, and different types of optical imaging, each with its own distinct advantages as a tool in the noninvasive, in vivo assessment of tumor angiogenesis (Glunde *et al.*, 2007). However, the assortment of available "functional" contrast mechanisms in conjunction with the development of novel imaging probes is making MRI a valuable imaging modality for the functional and molecular imaging of tumor vasculature in vivo. Here we have outlined the diversity and utility of MRI and provided brief descriptions of the biophysical phenomena underlying some of the contrast mechanisms used in MRI of tumor vasculature.

III. BASIS OF CONTRAST IN MR IMAGES

Almost every contrast mechanism for probing the tumor vasculature, including the use of exogenous MR contrast agents and targeted probes, in some way results from changes in the MR signal intensity brought about by changes in what are collectively known as tissue "relaxation times" (T_1 and T_2, or T_2^*). Briefly, the spin-lattice or longitudinal relaxation time (T_1) is the time constant that characterizes the exponential process by which the longitudinal component of the tissue magnetization returns or "relaxes" to its equilibrium value. It does so by exchanging energy with its surroundings or lattice, at the Larmor frequency ($\omega_0 = \gamma B_0$, where ω_0 is the rate of precession of the ensemble of spins under the influence of an applied magnetic field B_0 and γ is the gyromagnetic ratio). At the molecular scale, T_1 relaxation occurs through interactions between protons in tissue water and those on macromolecules or proteins, and by interactions with paramagnetic substances, that is, substances with unpaired electrons in their outermost shells. T_1-based MR contrast results from differences in T_1 dominating the signal intensity in the MR image. Tissues with short T_1s (such as fat) appear bright in T_1-weighted MRI, since the longitudinal magnetization recovers to equilibrium rapidly relative to tissues with long T_1s (such as cerebrospinal fluid).

Microscopic magnetic field heterogeneities in the applied magnetic field of an MRI scanner, as well as variations in local magnetic susceptibility due to the physiological microenvironment, cause spins contributing to the transverse component of the tissue magnetization, to lose phase coherence. The process through which this occurs is known as T_2^* relaxation. The loss in phase coherence attributable to static magnetic field heterogeneities can be recovered using a refocusing radiofrequency (RF) pulse or a spin-echo (SE) imaging method. However, as protons diffuse through the microscopic field inhomogeneities, they also lose phase coherence due to Brownian motion through these microscopic magnetic field gradients, which results in a phase dispersion that cannot be reversed by the application of a refocusing pulse. This is known as T_2 relaxation and in T_2-weighted MR images, tissues with short T_2s, such as the liver, appear dark due to the rapid decay of transverse magnetization compared to those with long T_2s, such as fat. Similarly, in T_2^*-weighted images, regions with large susceptibility gradients such as air–tissue interfaces of the orbits of the eye, or large veins carrying deoxygenated blood, appear hypointense. A computational model illustrating the T_2^* contrast mechanism is shown in Fig. 1.2 (Pathak et al., 2008b). Figure 1.2A is a volume rendering of a contrast agent bearing cerebrocortical capillary network from a rat brain, which results in a susceptibility gradient relative to the surrounding tissue. This susceptibility gradient in turn, perturbs the applied magnetic field at the microscopic scale as illustrated in Fig. 1.2B. Finally, water protons diffusing through these microscopic magnetic field heterogeneities experience different phase shifts resulting in a loss

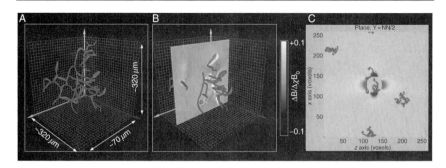

Figure 1.2. (A) 3D volume rendering of a rat cerebrocortical capillary network. (B) Capillary network in (A) overlaid with a slice through the 3D magnetic field map showing the microscopic magnetic field heterogeneities generated by the contrast agent bearing blood vessels. (C) Projection of five 3D proton diffusion trajectories onto a slice through the 3D magnetic field perturbation due to a contrast agent bearing blood vessel cylinder for the restricted proton diffusion case, that is, protons are not allowed to traverse the vessel boundary. These data are results of numerical simulations adapted with permission from Pathak *et al.* (2008b).

of phase coherence and attenuation of the MRI signal. An example of these diffusion trajectories is illustrated in Fig. 1.2C for the case of a single contrast agent bearing capillary.

A. Endogenous contrast

Probing tumor vasculature using the inherent or endogenous contrast produced by deoxyhemoglobin in tumor microvessels is based on the blood oxygenation level dependent (BOLD) contrast mechanism discovered by Ogawa (1990). The primary determinant of the MRI contrast observed is the concentration of paramagnetic deoxyhemoglobin. The presence of deoxyhemoglobin in a blood vessel causes a susceptibility difference between the microvessel and the surrounding tissue inducing microscopic magnetic field gradients that cause dephasing of the MR signal, leading to a reduction in the value of T_2^*. Since oxyhemoglobin is diamagnetic and does not produce the same dephasing, changes in blood oxygenation are observable as signal changes in T_2^*-weighted images. The dependence of T_2^* on oxygenation in a tissue can be approximated as

$$\frac{1}{T_2^*} \propto (1 - Y)b \qquad (1.1)$$

where Y is the fraction of oxygenated blood and b the fractional blood volume. In hypoxic tumors where $0 < Y < 0.2$, BOLD contrast is primarily dependent on b. This method works best in poorly oxygenated tumors and human xenografts with randomly oriented angiogenic microvessels. While this approach provides a fast and noninvasive measurement of tumor fractional blood volume without requiring the administration of an exogenous contrast, it does not provide quantitative assessment of tumor vascular volume, vascular permeability, or blood flow. Nonetheless, BOLD MRI has been used to detect changes in tumor oxygenation following induction of angiogenesis (Abramovitch et al., 1999), to obtain maps of the "functional" vasculature in genetically modified HIF-1 ($+/+$ and $-/-$) models (Carmeliet et al., 1998), and was recently shown to correlate with mean tumor pO2 measured using fluorine-19 [19F] MRI oximetry (Zhao et al., 2009). It is important to bear in mind that the relationship between angiogenesis and BOLD image contrast is complex as it is not solely determined by the oxygenation status of blood, but is also affected by factors such as oxygen saturation, hematocrit, blood flow, blood volume, vessel orientation, and geometry (Pathak et al., 2003).

In 2000, Silva et al. demonstrated the feasibility of imaging blood flow in a rodent brain tumor model, using another endogenous contrast MR mechanism known as arterial spin labeling (ASL) (Silva et al., 2000). In this approach, arterial blood water serves as the endogenous tracer, and is magnetically tagged proximal to the brain, using spatially selective RF inversion pulses. The effect of arterial tagging on downstream images can then be quantified in terms of tissue blood flow, since changes in signal intensity depend on the regional blood flow and degree of T_1 relaxation. Tissue blood flow images can then be computed from tagged and untagged control images. Although ASL exhibits sufficient sensitivity at high magnetic fields for mapping the tumor blood flow, it may not perform as well when blood flow in tumor vessels is very low, since the tagged arterial blood may not reach the tissue in time, relative to their T_1 relaxation time, that is, the spins will be fully relaxed by the time they enter the imaging volume. Recent studies have demonstrated excellent correlation between ASL and dynamic susceptibility contrast (DSC) MRI-derived measures of blood flow in patients with brain tumors, further establishing the utility and validity of ASL (Jarnum et al., 2010; Knutsson et al., 2010).

The advantage of characterizing the tumor vasculature with endogenous contrast-based MRI is that it does not require the intravenous (i.v.) administration of a contrast agent, making it entirely noninvasive, providing dynamic data with high temporal resolution. Repeated measurements in preclinical studies are therefore only limited by the constraints of anesthesia. However, using such approaches, one cannot quantify physiologic parameters like the tumor vascular volume or vascular permeability, which require the administration of exogenous MRI contrast agents.

B. Exogenous contrast

Unlike the dyes or tracers employed with nuclear medicine, X-ray or optical imaging techniques, MR contrast agents are not visualized directly in an MR image, but indirectly by the changes they induce in water proton relaxation behavior. The most commonly used MR contrast agents are paramagnetic gadolinium chelates. These agents are tightly bound complexes of the element gadolinium (Gd) and various chelating agents (Lauffer, 1987). The seven unpaired electrons of gadolinium produce a large magnetic moment that shortens both T_1 and T_2 of tissue water. Since tissue T_2 values are intrinsically shorter than the corresponding T_1 values, the T_1 effect of the contrast agent dominates, and tissues that take up the agent are brightened in T_1-weighted images. The magnetic susceptibility differences induced by Gd-based contrast agents shorten the T_2 and T_2^* of tissue water and tissues that take up the paramagnetic agent are darkened in T_2- and T_2^*-weighted images. A range of vascular parameters can be calculated from tracer kinetics principles, using the tissue concentration of Gd-based agents (Zaharchuk, 2007).

Gd-based complexes may be broadly classified as either being low molecular weight (≈ 0.57 kDa) agents, for example, gadolinium diethylenetriamine pentaacetic acid (GdDTPA) compounds used clinically for contrast enhancement of malignant tumors, or macromolecular agents (≈ 90 kDa) such as albumin-GdDTPA, that are confined to the vascular compartment for several hours and are used in preclinical studies of tumor angiogenesis. MR methods used to characterize tumor vascularization depend on the physical properties and pharmacokinetics of the contrast agent used (Fig. 1.3). A brief overview of some of these methods is presented in the subsequent sections.

C. Low molecular weight contrast agents

Several compounds in this class of paramagnetic agents are approved for routine clinical use and have been used to characterize the vasculature in a variety of tumors, including breast (Baar *et al.*, 2009), brain (Batchelor *et al.*, 2007), and cervical tumors (Mayr *et al.*, 2010). Most T_1 methods involve the analyses of relaxivity changes induced by the contrast agent to determine influx and outflux transfer constants, as well as the extracellular extravascular volume fraction based on one of several compartmental models (Fig. 1.3A; Tofts, 1997). Although not freely diffusible, this class of agent remains in the extracellular space, and three standard kinetic parameters can be derived from dynamic contrast-enhanced (DCE) T_1-weighted MRI: (i) K_{trans} (min^{-1}), which is the volume transfer constant between the blood plasma and the extravascular extracellular space (EES); (ii) k_{ep} (min^{-1}), which is the rate constant between the EES and blood plasma; and (iii) v_e (%), which is the volume of the EES per unit volume of tissue, that is, the volume fraction of the EES (Tofts, 1997). k_{ep} can be derived from the shape of the tracer concentration–time curve, but the determination of K_{trans} requires absolute values of the tracer concentration. K_{trans} has several different connotations depending on the balance between blood flow (F) and capillary permeability (P) in the tissue of interest. For example, in most

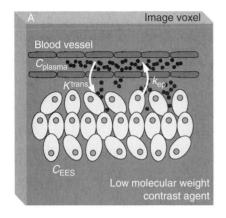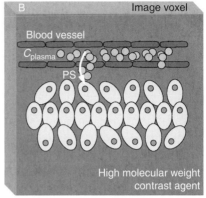

Figure 1.3. (A) Schematic of an imaging voxel relating the exchange of a low molecular weight contrast agent (e.g., GdDTPA) postadministration between plasma and the interstitial space. The plasma volume is usually small compared to the volume of the extravascular extracellular space or EES (shown in gray) and the intercellular space (shown in blue), which is inaccessible to the contrast agent. K_{trans} is the volume transfer constant between the blood plasma and EES, while kep is the rate constant between the EES and blood plasma. (B) Schematic of an imaging voxel demonstrating conditions for a macromolecular or high molecular weight agent (e.g., albumin-GdDTPA), which exhibits relatively slow leakage from the vasculature, longer circulation half-life, and equilibration of plasma concentrations within the tumor. Postadministration, the bulk of the T1 relaxation effect is proportional to the volume of the intravascular space since the contrast agent is initially confined to this space. As the contrast agent slowly extravasates into the EES, the volume transfer constant is proportional to the permeability surface area product (PS) of the tumor vasculature.

tumors the leakage concentration is limited by the flow-rate. Thus, in this flow-limited or high-permeability regime, solving the differential equation that relates tissue concentration (C_t) to the plasma concentration (C_p) yields: $K_{trans} = F\rho(1 - Hct)$, where F (ml g^{-1} min^{-1}) is the flow of whole blood per unit mass of tissue, ρ is the tissue density (g ml^{-1}), Hct the hematocrit, and $(1 - Hct)$ the plasma fraction. An excellent review of DCE MRI tracer kinetic models, terminology, and definitions can be found in Tofts *et al.* (1999). To calculate K_{trans} and $_{ve}$, the plasma and tissue concentrations of the contrast agent are required. Typically the plasma concentration (also known as the arterial input function, AIF) can be measured using voxels localized within a large blood vessel or approximated by a biexponential curve (Ohno *et al.*, 1979). In most DCE MRI models, the concentration of GdDTPA is measured from changes in the T_1 relaxation rate, assuming that water is in fast exchange between the vascular and extracellular compartments. It has been demonstrated by several investigators that the accuracy of

DCE MRI-derived vascular volumes can be significantly affected by incorrect assumptions regarding exchange rates (Donahue *et al.*, 1997; Kim *et al.*, 2002; Landis *et al.*, 2000).

As mentioned earlier, low molecular weight Gd-chelates employed in MRI produce both T_1 and T_2 relaxation effects. However, when high doses of these agents are employed, the induced bulk susceptibility differences between the intra- and extravascular spaces dominate classical dipole–dipole relaxation. MRI contrast can be engendered from local magnetic field heterogeneities either by proton diffusion through the microscopic field inhomogeneities or via intra-voxel dephasing. In the latter mechanism, the presence of microscopic field inhomogeneities within an imaging voxel produces a heterogeneity of resonant frequencies which affects the MR signal intensity even in the absence of diffusion. The effect of magnetic field inhomogeneities on transverse relaxation can be summarized as (Fisel *et al.*, 1991; Kennan, 1994):

$$\frac{1}{T_2^*} = \frac{1}{T_2} + \frac{1}{T'_2} \tag{1.2}$$

The relaxation rate $1/T_2^* (R_2^*)$ is the rate of free induction decay or the rate at which the gradient-echo (GE) amplitude decays. The relaxation rate $1/T_2(R2)$ is the rate at which the spin-echo (SE) amplitude decays. The relaxation rate, $1/T'_2(R'_2)$, is the water resonance line-width, which is a measure of the frequency distribution within a voxel. In the presence of a contrast agent bearing tumor vessel, the relative R2 and R_2^* relaxation rates depend on the diffusion coefficient (D) of spins in the vicinity of the induced magnetic field inhomogeneities, radius (R) of the tumor vessel, and the variation of the Larmor frequency ($d\omega$) at the surface of the vessel (Fisel *et al.*, 1991; Kennan, 1994; Weisskoff *et al.*, 1994; Yablonskiy, 1994). The relationship between the two physical characteristics (R and D) can be described in terms of the proton correlation time τ_D:

$$\tau_D = \frac{R^2}{D} \tag{1.3}$$

Depending on the relative magnitudes of $d\omega$ and τ_D, the magnitude of susceptibility-induced relaxation effects is divided into three regimes (Fisel *et al.*, 1991; Kennan, 1994; Villringer *et al.*, 1988; Weisskoff *et al.*, 1994; Yablonskiy, 1994).

(i) Fast exchange regime—in this regime, the rate of diffusion ($1/\tau_D$) is substantially greater than the frequency variation ($d\omega$), that is, $\tau_D\, d\omega \ll 1$. This is known as the "motionally narrowed" regime as the susceptibility-induced local magnetic field gradients are averaged out (Boxerman *et al.*, 1995; Fisel *et al.*, 1991; Kennan, 1994).

(ii) Slow exchange regime—in this regime, the rate of diffusion ($1/\tau_D$) is substantially smaller than the frequency variation ($d\omega$), that is, $\tau_D\,d\omega \gg 1$. Thus, the phase that a proton accumulates as it passes one perturber is large, that is, the effect is the same as it would be for the case of static field inhomogeneities. There is no signal attenuation on a T_2-weighted scan because the $180°$ RF pulse during the SE sequence refocuses static magnetic field inhomogeneities, while intravoxel dephasing still occurs in a GE sequence.

(iii) Intermediate exchange regime—in this regime, $\tau_D\,d\omega \sim 1$, that is, water diffusion is neither fast enough to be fully motionally narrowed nor slow enough to be approximated as linear gradients, making the description of the susceptibility-induced contrast more complex. In this regime, analytic solutions to estimate signal loss in the presence of diffusion becomes complicated due to the large spatial heterogeneity of the induced field gradients and numerical simulations are required (Boxerman et al., 1995; Kennan, 1994; Weisskoff et al., 1994).

A consequence of these regimes is that the SE and GE MRI methods have greatly different sensitivities to the size and scale of the field inhomogeneities, resulting in a differential sensitivity to contrast agent bearing tumor vessel caliber. It has been shown that the SE relaxation rate change (ΔR_2) increases, peaks for capillary-sized vessels (5–10 μm), and then decreases inversely with vessel radius, while the GE relaxation rate change (ΔR_2^*) increases and then plateaus to remain independent of vessel radius beyond capillary-sized vessels (Boxerman et al., 1995; Weisskoff et al., 1994). Therefore, SE methods are maximally sensitive to the microvascular blood volume, while the GE methods are more sensitive to the total blood volume. Based on this observation, SE sequences have been used in many tumor studies with the assumption that tumor angiogenesis is primarily characterized by an increase in the microvasculature (Aronen et al., 1993, 1994). However, given the large (> 20 μm) tortuous vessels usually found in tumors (Deane and Lantos, 1981; Pathak et al., 2001), whether SE or GE methods are most appropriate remains to be determined. Several investigators have acquired relative cerebral blood volume (rCBV) maps from first-pass DSC studies, with good spatio-temporal resolution (Maeda et al., 1993; Rosen et al., 1991). With this technique, preliminary results indicate that MRI-derived rCBV may better differentiate histologic tumor types than conventional MRI (Aronen et al., 1994) and provide information to predict tumor grade (Maeda et al., 1993; Schmainda et al., 2004). For a detailed description of susceptibility-based perfusion imaging of tumors see Pathak (2009).

Often MRI-derived concentration–time curves include contributions due to recirculation that must be eliminated before tumor blood volume and flow information can be extracted. This is usually accomplished by fitting to a γ-variate function with recirculation cut-off (Thompson et al., 1964). For an instantaneous bolus injection of MRI contrast agent, the central volume principle states that CBF=CBV/MTT, where MTT is the mean transit time of contrast agent through the vascular network (Zierler, 1962). However, bolus injections are of finite duration, and the measured concentration–time curve is the convolution of the ideal tissue–transit curve with the AIF. Thus, measurement of the blood flow requires knowledge of the arterial input curve to deconvolve the observed concentration–time curve (Axel, 1980; Ostergaard *et al.*, 1996).

Another complication with using first-pass techniques with low molecular weight Gd contrast agents is that with elevated permeability, as is often observed in tumor vasculature or with significant blood–brain barrier (BBB) disruption, as is often the case with brain tumors, contrast agent extravasates into the brain or tumor tissue resulting in enhanced T_1 relaxation effects. In such instances, signal increases due to T_1 effects can mask signal decreases due to T_2 or T_2^* effects, leading to an underestimation of rCBV. To address this issue, a method of analysis has been devised that corrects for these leakage effects when the leakage is not extreme (Donahue *et al.*, 2000). Another obstacle to the application of the central volume principle for the calculation of blood flow is the direct measurement of the MTT. Weisskoff *et al.* have demonstrated that MTT, which relates tissue blood volume to blood flow from the central volume principle, is not the first moment of the concentration–time curve for MR of intravascular tracers, and while first moment methods cannot be used by themselves to determine absolute flow, they do provide a useful relative measure of flow (Weisskoff, 1993).

The differential sensitivities of GE and SE methods to vessel radius have also been exploited to provide a measure of the average vessel size by measuring the ratio of GE and SE relaxation rates ($\Delta R_2^*/\Delta R_2$) (Dennie *et al.*, 1998). Donahue *et al.* (2000) demonstrated that clinically, the ratio $\Delta R_2^*/\Delta R_2$ correlated strongly with tumor grade and was a promising marker for the evaluation of tumor angiogenesis in patients.

All dynamic susceptibility-based contrast measurements are made assuming that the calibration factor that relates the relaxation rate to the contrast agent concentration is constant irrespective of tissue type. However, it has been shown that it is not the same for normal brain and tumor tissue and that this difference is due to the grossly different vascular morphology of tumors, due to tumor angiogenesis, compared to normal brain, and/or possibly differing blood rheological factors such as hematocrit (Pathak *et al.*, 2003, 2008b).

D. High molecular weight contrast agents

Quantitative assessment of tumor vascular parameters with low molecular weight contrast agents is often complicated by the rapid extravasation of the contrast agent from leaky tumor vessels. High molecular weight contrast agents, such as albumin-GdDTPA (alb-GdDTPA) complexes (Ogan *et al.*, 1987), or synthetic compounds, such as polylysine-GdDTPA (Weissleder *et al.*, 1998) and dendrimer Gd-complexes (Yu *et al.*, 2002), provide an opportunity for quantitative determination of tumor vascular volume and vascular permeability surface area product (PS). The relatively slow leakage of these agents from the vasculature results in longer circulation half-life and equilibration of plasma concentrations within the tumor. Assuming fast exchange of water between all the compartments in the tumor (plasma, interstitium, cells) the concentration of the contrast agent within any given voxel is proportional to changes in relaxation rate ($1/T_1$) before and after administration of the contrast agent. Relaxation rates can be measured either directly using fast single shot quantitative T_1 methods (Schwarzbauer *et al.*, 1993) or from T_1-weighted steady-state experiments (Brasch *et al.*, 1997), which provide better temporal resolution but are susceptible to experimental artifacts caused by variations in T_2 and T_2^* relaxation times. Pixel-wise maps can be generated from the acquired data and processed with the appropriate tracer kinetic model to obtain spatial maps of tumor vascular volume and vascular PS.

A simple linear compartment model, describing uptake of the contrast agent from plasma postulates a negligible reflux of the contrast agent from the interstitium back to the blood compartment (Fig. 1.3B). Blood concentrations of the contrast agent can be approximated to be constant for the duration of the MR experiment and under these conditions, contrast uptake can be modeled as a linear function of time (Patlak *et al.*, 1983; Roberts *et al.*, 2000). In this case, the slope of the contrast agent concentration versus time plot provides the vascular PS, and the y-intercept provides the vascular volume. For absolute values of these parameters, the change in relaxation rate of the blood must be quantified, which can be obtained separately from blood samples taken before injection of the contrast agent and at the end of the experiment, or may be obtained noninvasively (Pathak *et al.*, 2004).

Tissue concentrations of the agent over a period of up to 40 min after the bolus injection increase linearly with time. Therefore, the simple linear model is preferable for analysis of intrinsically noisy relaxation data as it is much more stable in comparison with nonlinear fitting algorithms required for the two compartment models discussed above. This linear-model approach has been employed to detect vascular differences for metastatic versus nonmetastatic breast and prostate cancer xenografts (Bhujwalla *et al.*, 2001). The accuracy of the measurement of tissue vascular volume using this approach does, however, depends on the water exchange rate between the vascular and extracellular compartments. As mentioned earlier, using a simplified model of fast exchange where there may be intermediate to slow exchange can lead to significant underestimation of vascular volume. Experimental approaches to minimize these errors are based on observations that the initial slope of the relaxation curve is independent of the exchange rate (Donahue *et al.*, 1994, 1996). Large molecular weight contrast agents have also been used to measure tumor blood flow by detecting the first pass of the agent through tumor vasculature, similar to the method described in Ostergaard *et al.* (1996), although this approach may not be feasible when the heartbeat is very rapid as in preclinical studies with rodents.

IV. IMAGING RECEPTOR EXPRESSION

The development and availability of contrast agents that generate receptor or molecular-targeted contrast has resulted in exciting new capabilities to characterize tumor vasculature and vascular targets with MRI. Mechanisms for generating MR contrast include the use of paramagnetic and superparamagnetic agents. Targeted contrast agents can be directed to cell-surface receptors expressed on tumor endothelial cells using peptides, ligands, or antibodies. These molecular imaging capabilities in combination with the strong functional imaging capabilities of MR methods will allow molecular-functional characterization of tumor vasculature.

Most of the receptor imaging studies in MRI studies of tumor angiogenesis have used imaging probes that bind to integrins such as $\alpha_v\beta_3$ (Sipkins *et al.*, 1998) and $\alpha_5\beta_1$ (Schmieder *et al.*, 2008) that are expressed on the surface of endothelial cells during neovascularization. MR contrast can be generated to image tumor angiogenesis using probes that bind to these integrins. However, MRI studies of tumor angiogenesis using targeted contrast agents are currently at the preclinical stage. A major goal for the future is to translate these preclinical developments to the clinic. The inherent low sensitivity of molecular MRI methods requires higher concentrations of MR contrast agents compared to radiopharmaceuticals and is therefore subject to more stringent Food and Drug Administration (FDA) control. The risk of nephrogenic systemic fibrosis arising from gadolinium-based contrast agents administrated to patients with kidney disease is a major concern (Neuwelt *et al.*, 2009). Recent studies have therefore focused on finding alternative MR contrast agents, such as ultrasmall superparamagnetic iron oxide (USPIO) particles, for patients at risk for nephrogenic systemic fibrosis (Neuwelt *et al.*, 2009). Unlike delivery of conventional contrast agents within the tumor interstitium where its size limits delivery, the advantage of vascular molecular imaging is that the size of the probe does not pose a delivery problem since the receptors or targets are directly accessible to the probes circulating in the vasculature. In fact, larger sizes prolong the circulation time of the probes within the vasculature, resulting in better pharmacokinetics.

Since the RGD (Arg-Gly-Asp) peptide specifically binds to the $\alpha_v\beta_3$ integrin, USPIOs were combined with RGD peptides to visualize integrin expression through the T_2 contrast generated by the RGD–USPIOs (Zhang *et al.*, 2007).

The contrast generated was found to depend on the level of $\alpha_v\beta_3$ integrin expression in different xenograft models. More recently, superparamagnetic polymeric micelles (SPPM), which consist of a cluster of SPIOs within a hydrophobic core, were used in combination with an off-resonance saturation (ORS) imaging sequence to detect $\alpha_v\beta_3$ integrin expression, as shown in Fig. 1.4. The combination of these sensitive detection probes (~ 70 nm in diameter) together with the ORS imaging allowed target detection within the picomolar range (Khemtong et al., 2009).

Dual $\alpha_5\beta_1(\alpha_v\beta_3)$-targeted paramagnetic nanoparticles incorporating the antiangiogenic agent fumagillin were also used to create a "theranostic" agent to demonstrate the antiangiogenic effect of fumagillin with 3D MRI in preclinical studies (Schmieder et al., 2008). Another colloidal iron oxide nanoparticle (CION) was also recently reported that consists of oleate-coated magnetite particles within a cross-linked phospholipid nanoemulsion (Senpan et al., 2009). The sensitivity of detection reported for CION was within the nanomolar range. The outer phospholipid monolayer of the CION can be used to entrap drugs for targeted delivery to cells following membrane hemifusion. Challenges for the future will be to continue to increase the sensitivity of detection of such agents through novel chemistry, or signal amplification strategies such as hyperpolarization (Ardenkjaer-Larsen et al., 2003; Day et al., 2007), or instrumentation and the incorporation of multimodality imaging.

V. IMAGING VASCULAR TARGETING

Most MRI studies of vascular targeting have used functional parameters to detect changes in vasculature following the targeting. However, the great promise of noninvasive modalities such as MRI with their breadth of functional imaging capabilities is to combine molecular imaging with functional characterization, noninvasively and in vivo. The need to evaluate the efficacy of antiangiogenic and antivascular agents in vivo has created a major demand for noninvasive imaging biomarkers to individualize treatments, both in terms of matching the most suitable treatment to the patient and to determine the response to treatment (Jain et al., 2009; Sorensen et al., 2009). Most antiangiogenic treatments are combined with standard chemotherapy, creating a further challenge for identifying suitable noninvasive imaging biomarkers to detect the effectiveness of the antiangiogenic treatment.

Several preclinical studies have identified changes in DCE-derived parameters following antiangiogenic treatment. For example, iron oxide nanoparticle based studies have detected early vascular changes, as detected by a significant decrease of vascular volume following antiangiogenic treatment with VEGF (vascular endothelial growth factor) targeting (JuanYin et al., 2009;

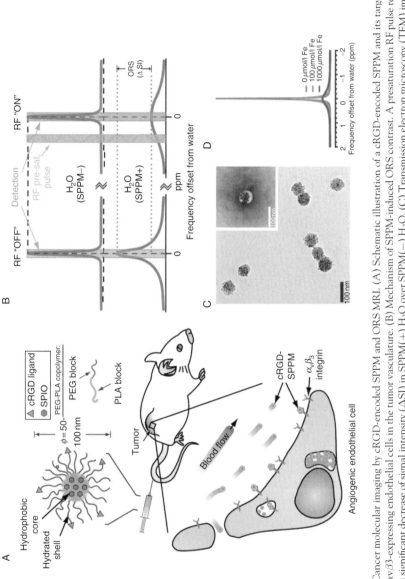

Figure 1.4. Cancer molecular imaging by cRGD-encoded SPPM and ORS MRI. (A) Schematic illustration of a cRGD-encoded SPPM and its targeting to $\alpha v \beta 3$-expressing endothelial cells in the tumor vasculature. (B) Mechanism of SPPM-induced ORS contrast. A presaturation RF pulse results in a significant decrease of signal intensity (ΔSI) in SPPM($-$) H$_2$O over SPPM($+$) H$_2$O. (C) Transmission electron microscopy (TEM) image of a representative SPPM sample. Inset, a SPPM particle after negative staining with 2% phosphotungstic acid (PTA) solution. (D) 1H NMR (300 MHz) spectra of water containing different concentrations of SPPM (in [Fe]/mmol/L). Adapted with permission from Khemtong *et al.* (2009). PEG-PLA, poly(ethylene glycol)-block-poly(d,l-lactide).

Varallyay et al., 2009). USPIOs were also used to demonstrate the antiangiogenic effects of a selective thrombogenic vascular-targeting agent consisting of the fusion peptide tTF-RGD that contains the thrombosis-inducing truncated tissue factor (tTF) together with the tumor vascular-targeting RGD peptide (Persigehl et al., 2007).

Gd-based agents have also been used to assess early vascular changes following antiangiogenic treatment including VEGF targeting. Treatment with bevacizumab significantly decreased tumor vascular permeability as measured by the low molecular weight contrast agent gadodiamide (Varallyay et al., 2009). The spatial heterogeneity of response to a VEGF-receptor tyrosine kinase inhibitor was evident in a study using both low and high molecular weight Gd-based contrast agents (Li et al., 2005).

DCE MRI has also been used in the clinic to determine changes following antiangiogenic treatment. Most of these clinical studies have focused on the inhibition of VEGF using bevacizumab for a range of tumors such as glioblastomas (Batchelor et al., 2007), rectal cancer (Willett et al., 2009), and hepatocellular carcinoma (Zhu et al., 2008). While a decrease of DCE-parameters was detected following treatment, additional studies are required to develop DCE MRI parameters as biomarkers to predict and detect response to antiangiogenic treatment (Jain et al., 2009).

In addition to antiangiogenic agents, MRI is useful for detecting the effects of standard chemotherapeutic agents as well as novel molecular targeting agents on tumor vasculature. An increase in tumor permeability was detected using DCE MRI following TGF-β type 1 receptor inhibition (Minowa et al., 2009). In this study, liposomal-GdDTPA ($\sim$120 nm size) was used as a macromolecular contrast agent that reported increased permeability following TGF-β inhibition. Liposomes are also used as carriers of anticancer drugs or molecular-based therapeutic agents such as small interfering RNA (siRNA). Decorating the liposomal membrane with MRI reporters has also been used to visualize the delivery of these carriers containing siRNA to tumors (Mikhaylova et al., 2009). Image-guided tumor vascular specific delivery of MRI-detectable liposomes carrying therapeutic cargo as "theranostic agents" may be achieved by targeting the liposomes using peptides or antibodies that are specific to tumor vasculature (Torchilin, 2005).

VI. MULTIMODAL MOLECULAR-FUNCTIONAL IMAGING OF TUMOR VASCULATURE

The ultimate goal of noninvasive imaging is to be able to image a specific genetic or molecular alteration and its effects on downstream function. However, the restrictions imposed by the low sensitivity of MR detection of contrast agents ($\sim$0.1 mM) limit the detection of receptors and molecular targets that are present at low concentrations (Artemov et al., 2004). Therefore, achieving molecular-functional imaging capabilities will require combining the strengths of MRI with the high sensitivity of complementary imaging modalities such as optical imaging for preclinical studies, and nuclear imaging for clinical applications. Indeed, several examples of preclinical and clinical research studies are providing exciting glimpses of the promise of multimodality imaging in understanding, characterizing, and targeting tumor angiogenesis. Advances in multimodality imaging are occurring at the level of novel multimodal probe development as well in the development of new imaging instrumentation capable of acquiring colocalized multimodal images. Several studies have already incorporated image segmentation and the landmark-based image fusion for images acquired from separate modalities. Image fusion of $[_{18}F]$ Galacto-RGD PET images with MRI using anatomical landmarks revealed good correlation between immunohistochemical distribution of $\alpha_v\beta_3$ expression and PET image intensity in human squamous cell carcinoma of the head and neck (Beer et al., 2007). Regions of increased contrast uptake were identified in tumor subvolumes, demonstrating the feasibility of relating molecular expression to function with combined PET/MRI (Fig. 1.5).

In a preclinical study, VEGFR2 targeted microbubbles were used for contrast-enhanced ultrasound and correlated with DCE MRI parameters. While no discernible patterns between VEGFR2 expression and K_{trans} were evident, these studies demonstrate the feasibility of molecular-functional vascular imaging (Loveless et al., 2009). Multimodal probes that can be used in combination with two or more imaging modalities are also being developed. Recently high-resolution SPECT–CT/MRI of angiogenesis in a Vx2 rabbit tumor model was reported using probes that consisted of $\alpha_v\beta_3$-targeted 99mTc nanoparticles (Lijowski et al., 2009).

MRI can also be integrated with magnetic resonance spectroscopy (MRS) and magnetic resonance spectroscopic imaging (MRSI) to understand the relationship between tumor vasculature and pH or metabolism (Bhujwalla et al., 2002; Provent et al., 2007). In preclinical studies, multimodal approaches combining optical imaging of reporters driven by specific molecular pathways or conditions can be integrated with MRI to understand the relationship between molecular conditions such as hypoxia, or specific pathways and tumor angiogenesis and vascularization. We have used combined MRI, MRSI, and optical imaging, to study a prostate cancer model, which expresses enhanced green fluorescent protein (EGFP) under hypoxia (Raman et al., 2006). In this study, a multiparametric approach of combined vascular and optical imaging was used to obtain vascular and hypoxia maps, from colocalized regions within a tumor (Raman et al., 2006). In this tumor model, hypoxic regions were found to exhibit low vascular volume and high permeability. More recently a combined vascular MRI, metabolic MRSI, and optical characterization of the same tumor model identified a combination of vascular, metabolic, and hypoxic conditions that characterize metastasis-permissive environments (Penet et al., 2009).

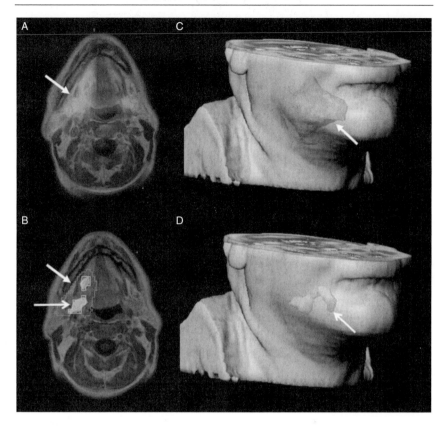

Figure 1.5. Patient with a squamous cell carcinoma of the head and neck in the right oral cavity. The [^{18}F]Galacto-RGD PET/MRI image fusion (A) shows intense and heterogeneous tracer uptake in the lesion (arrow). Moderate uptake is also notable in the submandibular gland. A transaxial MRI slice of the tumor volume as defined by MRI (B, closed-tipped arrow; red line) and in the corresponding 3D reconstruction (C, closed-tipped arrow). By applying a threshold of standardized uptake values (SUV) = 3 and only using pixels with SUVs above this threshold, a subvolume with more intense $\alpha_v\beta_3$ expression could be defined (blue line and blue area, B; open-tipped arrow), (D, blue volume; open-tipped arrow). Adapted with permission from (Beer *et al.*, 2007). (See Color Insert.)

VII. CONCLUSION

MRI is continuing to evolve as an important modality for the molecular-functional characterization of the tumor vasculature. Despite the initial promise of antiangiogenic agents to effectively treat cancer, in reality, the success of these agents has been limited, due in part to the inability to stratify patients for the appropriate targeted therapies. As noninvasive molecular imaging methods develop, these will play an important role in patient selection and "personalization" of therapy.

Stem cells and endothelial progenitor cells are increasingly being implicated in tumor vascularization and the failure of antiangiogenic treatment (Monzani and La Porta, 2008; van der Schaft et al., 2004). Imaging of MRI reporter-labeled stem cells or endothelial progenitor cells will provide new insights into their role in tumor angiogenesis.

Major advances are also being made in the identification of tumor vascular specific receptors (Lewis et al., 2009) and through phage display (Driessen et al., 2009; Kolonin et al., 2001). The availability of highly specific tumor vascular markers opens avenues for image-guided targeting of the tumor angiogenesis using nanodevices that deliver therapeutic agents, gene delivery, or prodrug enzymes. With exciting developments in siRNA technology, the image-guided delivery of siRNA to tumor vasculature, visualization of this delivery via liposome technology or other nanocarriers, and detection of a therapeutic response are well within the realm of current imaging capabilities. A major direction for the future will also be to translate preclinical discoveries for use in the clinic. The availability of multimodality imaging systems that combine the strengths and capabilities of each modality should facilitate this translation.

Acknowledgments

Support from P50 CA103175, R01 CA73850, R01 CA82337, R01 CA136576, R01 CA138515, R01 CA138264, R21 CA140904, R21 CA133600, R21 CA128793 (APP), and KG090640 (APP) is gratefully acknowledged.

References

Abramovitch, R., Dafni, H., Smouha, E., Benjamin, L., and Neeman, M. (1999). In vivo prediction of vascular susceptibility to vascular endothelial growth factor withdrawal: Magnetic resonance imaging of C6 rat glioma in nude mice. Cancer Res. 59, 5012–5016.

Ardenkjaer-Larsen, J. H., Fridlund, B., Gram, A., Hansson, G., Hansson, L., Lerche, M. H., Servin, R., Thaning, M., and Golman, K. (2003). Increase in signal-to-noise ratio of >10,000 times in liquid-state NMR. Proc. Natl. Acad. Sci. USA 100(18), 10158–10163.

Aronen, H. J., Cohen, M. S., Belliveau, J. W., Fordham, J. A., and Rosen, B. R. (1993). Ultrafast imaging of brain tumors. Top. Magn. Reson. Imaging 5(1), 14–24.

Aronen, H. J., Gazit, I. E., Louis, D. N., Buchbinder, B. R., Pardo, F. S., Weisskoff, R. M., Harsh, G. R., Cosgrove, G. R., Halpern, E. F., Hochberg, F. H., and Rosen, B. R. (1994). Cerebral blood volume maps of gliomas: Comparison with tumor grade and histological findings. Radiology 191, 41–51.

Artemov, D., Bhujwalla, Z. M., and Bulte, J. W. (2004). Magnetic resonance imaging of cell surface receptors using targeted contrast agents. Curr. Pharm. Biotechnol. 5(6), 485–494.

Axel, L. (1980). Cerebral blood flow determination by rapid-sequence computed tomography: Theoretical analysis. *Radiology* **137**(3), 679–686.

Baar, J., Silverman, P., Lyons, J., Fu, P., Abdul-Karim, F., Ziats, N., Wasman, J., Hartman, P., Jesberger, J., Dumadag, L., Hohler, E., Leeming, R., *et al.* (2009). A vasculature-targeting regimen of preoperative docetaxel with or without bevacizumab for locally advanced breast cancer: impact on angiogenic biomarkers. *Clin. Cancer Res.* **15**(10), 3583–3590.

Baluk, P., Hashizume, H., and McDonald, D. M. (2005). Cellular abnormalities of blood vessels as targets in cancer. *Curr. Opin. Genet. Dev.* **15**(1), 102–111.

Batchelor, T. T., Sorensen, A. G., di Tomaso, E., Zhang, W. T., Duda, D. G., Cohen, K. S., Kozak, K. R., Cahill, D. P., Chen, P. J., Zhu, M., Ancukiewicz, M., Mrugala, M. M., *et al.* (2007). AZD2171, a pan-VEGF receptor tyrosine kinase inhibitor, normalizes tumor vasculature and alleviates edema in glioblastoma patients. *Cancer Cell* **11**(1), 83–95.

Beer, A. J., Grosu, A. L., Carlsen, J., Kolk, A., Sarbia, M., Stangier, I., Watzlowik, P., Wester, H. J., Haubner, R., and Schwaiger, M. (2007). [^{18}F]galacto-RGD positron emission tomography for imaging of alphavbeta3 expression on the neovasculature in patients with squamous cell carcinoma of the head and neck. *Clin. Cancer Res.* **13**(22 Pt 1), 6610–6616.

Bertout, J. A., Patel, S. A., and Simon, M. C. (2008). The impact of O2 availability on human cancer. *Nat. Rev. Cancer* **8**(12), 967–975.

Bhujwalla, Z. M., Artemov, D., Ballesteros, P., Cerdan, S., Gillies, R. J., and Solaiyappan, M. (2002). Combined vascular and extracellular pH imaging of solid tumors. *NMR Biomed* **15**(2), 114–119.

Bhujwalla, Z. M., Artemov, D., Natarajan, K., Ackerstaff, E., and Solaiyappan, M. (2001). Vascular differences detected by MRI for metastatic versus nonmetastatic breast and prostate cancer xenografts. *Neoplasia* **3**(2), 143–153.

Boxerman, J. L., Hamberg, L. M., Rosen, B. R., and Weisskoff, R. M. (1995). MR contrast due to intravascular magnetic susceptibility perturbations. *Magn. Reson. Med.* **34**, 555–566.

Brasch, R., Pham, C., Shames, D., *et al.* (1997). Assessing tumor angiogenesis using macromolecular MR imaging contrast media. *J. Magn. Reson. Imaging* **7**, 68–74.

Brooks, P. C., Montgomery, A. M., Rosenfeld, M., Reisfeld, R. A., Hu, T., Klier, G., and Cheresh, D. A. (1994). Integrin alpha v beta 3 antagonists promote tumor regression by inducing apoptosis of angiogenic blood vessels. *Cell* **79**(7), 1157–1164.

Brown, L. F., Berse, B., Jackman, R. W., Tognazzi, K., Manseau, E. J., Dvorak, H. F., and Senger, D. R. (1993). Increased expression of vascular permeability factor (vascular endothelial growth factor) and its receptors in kidney and bladder carcinomas. *Am. J. Pathol.* **143**(5), 1255–1262.

Burrows, F. J., Derbyshire, E. J., Tazzari, P. L., Amlot, P., Gazdar, A. F., King, S. W., Letarte, M., Vitetta, E. S., and Thorpe, P. E. (1995). Up-regulation of endoglin on vascular endothelial cells in human solid tumors: Implications for diagnosis and therapy. *Clin. Cancer Res.* **1**(12), 1623–1634.

Carmeliet, P. (2005). Angiogenesis in life, disease and medicine. *Nature* **438**(7070), 932–936.

Carmeliet, P., Dor, Y., Herbert, J.-M., Fukumura, D., Brusselmans, K., Dewerchin, M., Neeman, M., Bono, F., Abramovitch, R., Maxwell, P., Koch, C. J., Ratcliffe, P., *et al.* (1998). Role of HIF-1 in hypoxia-mediated apoptosis, cell proliferation and tumor angiogenesis. *Nature* **394**, 485–490.

Chang, Y. S., di Tomaso, E., McDonald, D. M., Jones, R., Jain, R. K., and Munn, L. L. (2000). Mosaic blood vessels in tumors: Frequency of cancer cells in contact with flowing blood. *Proc. Natl. Acad. Sci. USA* **97**(26), 14608–14613.

Clark, E. R., Hitschler, W. J., KirbySmith, H. T., Rex, R. O., and Smith, J. H. (1931). General observations on the ingrowth of new blood vessels into standardized chambers in the rabbit's ear, and the subsequent changes in the newly grown vessels over a period of months. *Anat. Rec.* **50**, 129–167.

Corada, M., Zanetta, L., Orsenigo, F., Breviario, F., Lampugnani, M. G., Bernasconi, S., Liao, F., Hicklin, D. J., Bohlen, P., and Dejana, E. (2002). A monoclonal antibody to vascular endothelial-cadherin inhibits tumor angiogenesis without side effects on endothelial permeability. *Blood* **100**(3), 905–911.

Day, S. E., Kettunen, M. I., Gallagher, F. A., Hu, D. E., Lerche, M., Wolber, J., Golman, K., Ardenkjaer-Larsen, J. H., and Brindle, K. M. (2007). Detecting tumor response to treatment using hyperpolarized 13C magnetic resonance imaging and spectroscopy. *Nat. Med.* **13**(11), 1382–1387.

Deane, B. R., and Lantos, P. L. (1981). The vasculature of experimental brain tumors–Part 1: A sequential light and electron microscope study of angiogenesis. *J. Neurol. Sci.* **49**, 55–66.

Dennie, J., Mandeville, J. B., Boxerman, J. L., Packard, S. D., Rosen, B. R., and Weisskoff, R. M. (1998). NMR imaging of changes in vascular morphology due to tumor angiogenesis. *Magn. Reson. Med.* **40**, 793–799.

Donahue, K. M., Burstein, D., MAnning, W. J., and Gray, M. L. (1994). Studies of Gd-DTPA relaxivity and proton exchange rates in tissue. *Magn. Reson. Med.* **32**, 66–75.

Donahue, K. M., Krouwer, H. G., Rand, S. D., Pathak, A. P., Marszalkowski, C. S., Censky, S. C., and Prost, R. W. (2000). Utility of simultaneously acquired gradient-echo and spin-echo cerebral blood volume and morphology maps in brain tumor patients. *Magn. Reson. Med.* **43**(6), 845–853.

Donahue, K. M., Weisskoff, R. M., and Burstein, D. (1997). Water diffusion and exchange as they influence contrast enhancement. *J. Magn. Reson. Imaging* **7**(1), 102–110.

Donahue, K. M., Weisskoff, R. M., Chesler, D. A., Kwong, K. K., Bogdanov, A. A., Mandeville, J. B., and Rosen, B. R. (1996). Improving MR quantitation of regional blood volume with intravascular T_1 contrast agents: Accuracy, precision, and water exchange. *Magn. Reson. Med.* **36**, 858–867.

Driessen, W. H., Ozawa, M. G., Arap, W., and Pasqualini, R. (2009). Ligand-directed cancer gene therapy to angiogenic vasculature. *Adv. Genet.* **67**, 103–121.

Epstein, A. L., Khawli, L. A., Hornick, J. L., and Taylor, C. R. (1995). Identification of a monoclonal antibody, TV-1, directed against the basement membrane of tumor vessels, and its use to enhance the delivery of macromolecules to tumors after conjugation with interleukin 2. *Cancer Res.* **55**(12), 2673–2680.

Fisel, C. R., Ackerman, J. L., Buxton, R. B., Garrido, L., Belliveau, J. W., Rosen, R. B., and Brady, T. J. (1991). MR contrast due to microscopically heterogeneous magnetic susceptibility: Numerical simulations and applications to cerebral physiology. *Magn. Reson. Med.* **17**, 336–347.

Folkman, J. (1971). Tumor angiogenesis: therapeutic implications. *N. Engl. J. Med.* **285**(21), 1182–1186.

Folkman, J. (2007). Angiogenesis: An organizing principle for drug discovery? *Nat. Rev. Drug Discov.* **6**(4), 273–286.

Forster-Horvath, C., Meszaros, L., Raso, E., Dome, B., Ladanyi, A., Morini, M., Albini, A., and Timar, J. (2004). Expression of CD44v3 protein in human endothelial cells in vitro and in tumoral microvessels in vivo. *Microvasc. Res.* **68**(2), 110–118.

Glunde, K., Pathak, A. P., and Bhujwalla, Z. M. (2007). Molecular-functional imaging of cancer: to image and imagine. *Trends Mol. Med.* **13**(7), 287–297.

Goldman, E. (1907). The growth of malignant disease in man and the lower animals, with special reference to the vascular system. *Proc. R. Soc. Med.* **1**, 1–13.

Hagemeier, H. H., Vollmer, E., Goerdt, S., Schulze-Osthoff, K., and Sorg, C. (1986). A monoclonal antibody reacting with endothelial cells of budding vessels in tumors and inflammatory tissues, and non-reactive with normal adult tissues. *Int. J. Cancer* **38**(4), 481–488.

Hallahan, D. E., Staba-Hogan, M. J., Virudachalam, S., and Kolchinsky, A. (1998). X-ray-induced P-selectin localization to the lumen of tumor blood vessels. *Cancer Res.* **58**(22), 5216–5220.

Holash, J., Maisonpierre, P. C., Compton, D., Boland, P., Alexander, C. R., Zagzag, D., Yancopoulos, G. D., and Wiegand, S. J. (1999). Vessel cooption, regression, and growth in tumors mediated by angiopoietins and VEGF. *Science* **284**(5422), 1994–1998.

Hunter, J. (1794). Treatise on the blood, inflammation and gunshot wounds. Thomas Bradford, Phildelphia.

Jain, R. K. (1988). Determinants of tumor blood flow: A review. *Cancer Res.* **48**(10), 2641–2658.

Jain, R. K., and Carmeliet, P. F. (2001). Vessels of death or life. *Sci. Am.* **285**(6), 38–45.

Jain, R. K., Duda, D. G., Willett, C. G., Sahani, D. V., Zhu, A. X., Loeffler, J. S., Batchelor, T. T., and Sorensen, A. G. (2009). Biomarkers of response and resistance to antiangiogenic therapy. *Nat. Rev. Clin. Oncol.* **6**(6), 327–338.

Jarnum, H., Steffensen, E. G., Knutsson, L., Frund, E. T., Simonsen, C. W., Lundbye-Christensen, S., Shankaranarayanan, A., Alsop, D. C., Jensen, F. T., and Larsson, E. M. (2010). Perfusion MRI of brain tumours: a comparative study of pseudo-continuous arterial spin labelling and dynamic susceptibility contrast imaging. *Neuroradiology* **52**(4), 307–317.

JuanYin, J., Tracy, K., Zhang, L., Munasinghe, J., Shapiro, E., Koretsky, A., and Kelly, K. (2009). Noninvasive imaging of the functional effects of anti-VEGF therapy on tumor cell extravasation and regional blood volume in an experimental brain metastasis model. *Clin. Exp. Metastasis* **26**(5), 403–414.

Kalluri, R. (2003). Basement membranes: Structure, assembly and role in tumour angiogenesis. *Nat. Rev. Cancer* **3**(6), 422–433.

Karumanchi, S. A., Jha, V., Ramchandran, R., Karihaloo, A., Tsiokas, L., Chan, B., Dhanabal, M., Hanai, J. I., Venkataraman, G., Shriver, Z., Keiser, N., Kalluri, R., *et al.* (2001). Cell surface glypicans are low-affinity endostatin receptors. *Mol. Cell* **7**(4), 811–822.

Kennan, R. P. (1994). Intravascular susceptibility contrast mechanisms in tissues. *Magn. Reson. Med.* **31**, 9–21.

Khemtong, C., Kessinger, C. W., Ren, J., Bey, E. A., Yang, S. G., Guthi, J. S., Boothman, D. A., Sherry, A. D., and Gao, J. (2009). In vivo off-resonance saturation magnetic resonance imaging of alphavbeta3-targeted superparamagnetic nanoparticles. *Cancer Res.* **69**(4), 1651–1658.

Kim, Y. R., Rebro, K. J., and Schmainda, K. M. (2002). Water exchange and inflow affect the accuracy of T_1-GRE blood volume measurements: Implications for the evaluation of tumor angiogenesis. *Magn. Reson. Med.* **47**(6), 1110–1120.

Knutsson, L., van Westen, D., Petersen, E. T., Bloch, K. M., Holtas, S., Stahlberg, F., and Wirestam, R. (2010). Absolute quantification of cerebral blood flow: correlation between dynamic susceptibility contrast MRI and model-free arterial spin labeling. *Magn. Reson. Imaging* **28**(1), 1–7.

Koivunen, E., Arap, W., Valtanen, H., Rainisalo, A., Medina, O. P., Heikkila, P., Kantor, C., Gahmberg, C. G., Salo, T., Konttinen, Y. T., Sorsa, T., Ruoslahti, E., *et al.* (1999). Tumor targeting with a selective gelatinase inhibitor. *Nat. Biotechnol.* **17**(8), 768–774.

Kolonin, M., Pasqualini, R., and Arap, W. (2001). Molecular addresses in blood vessels as targets for therapy. *Curr. Opin. Chem. Biol.* **5**(3), 308–313.

Konerding, M. A., van Ackern, C., Fait, E., Steinberg, F., and Streffer, C. (2000). Morphological aspects of tumor angiogenesis and microcirculation. Blood Perfusion and Microenvironment of Human Tumors: Implications for Clinical Radiooncology. Springer. Verlag, New York, pp. 5–17.

Landis, C. S., Li, X., Telang, F. W., Coderre, J. A., Micca, P. L., Rooney, W. D., Latour, L. L., Vetek, G., Palyka, I., and Springer, C. S., Jr. (2000). Determination of the MRI contrast agent concentration time course in vivo following bolus injection: Effect of equilibrium transcytolemmal water exchange. *Magn. Reson. Med.* **44**(4), 563–574.

Langenkamp, E., and Molema, G. (2009). Microvascular endothelial cell heterogeneity: General concepts and pharmacological consequences for anti-angiogenic therapy of cancer. *Cell Tissue Res.* **335**(1), 205–222.

Lauffer, R. B. (1987). Paramagnetic metal complexes as water proton relaxation agents for NMR imaging: Theory and design. *Chem. Rev.* **87**, 901–927.

Lewis, V. O., Ozawa, M. G., Deavers, M. T., Wang, G., Shintani, T., Arap, W., and Pasqualini, R. (2009). The interleukin-11 receptor alpha as a candidate ligand-directed target in osteosarcoma: Consistent data from cell lines, orthotopic models, and human tumor samples. *Cancer Res.* **69**(5), 1995–1999.

Lewis, W. (1927). The vascular patterns of tumors. *Bull. Johns Hopkins Hosp.* **41,** 156–162.

Li, K. L., Wilmes, L. J., Henry, R. G., Pallavicini, M. G., Park, J. W., Hu-Lowe, D. D., McShane, T. M., Shalinsky, D. R., Fu, Y. J., Brasch, R. C., and Hylton, N. M. (2005). Heterogeneity in the angiogenic response of a BT474 human breast cancer to a novel vascular endothelial growth factor-receptor tyrosine kinase inhibitor: Assessment by voxel analysis of dynamic contrast-enhanced MRI. *J. Magn. Reson. Imaging* **22**(4), 511–519.

Lijowski, M., Caruthers, S., Hu, G., Zhang, H., Scott, M. J., Williams, T., Erpelding, T., Schmieder, A. H., Kiefer, G., Gulyas, G., Athey, P. S., Gaffney, P. J., *et al.* (2009). High sensitivity: High-resolution SPECT-CT/MR molecular imaging of angiogenesis in the Vx2 model. *Invest. Radiol.* **44**(1), 15–22.

Loveless, M. E., Whisenant, J. G., Wilson, K., Lyshchik, A., Sinha, T. K., Gore, J. C., and Yankeelov, T. E. (2009). Coregistration of ultrasonography and magnetic resonance imaging with a preliminary investigation of the spatial colocalization of vascular endothelial growth factor receptor 2 expression and tumor perfusion in a murine tumor model. *Mol. Imaging* **8**(4), 187–198.

Maeda, M., Itoh, S., Kimura, H., Iwasaki, T., Hayashi, N., Yamamoto, K., Ishii, Y., and Kubota, T. (1993). Tumor vascularity in the brain: Evaluation with dynamic susceptibility-contrast MR imaging. *Radiology* **189,** 233–238.

Maniotis, A. J., Folberg, R., Hess, A., Seftor, E. A., Gardner, L. M., Pe'er, J., Trent, J. M., Meltzer, P. S., and Hendrix, M. J. (1999). Vascular channel formation by human melanoma cells in vivo and in vitro: Vasculogenic mimicry. *Am. J. Pathol.* **155**(3), 739–752.

Mayr, N. A., Yuh, W. T., Jajoura, D., Wang, J. Z., Lo, S. S., Montebello, J. F., Porter, K., Zhang, D., McMeekin, D. S., and Buatti, J. M. (2010). Ultra-early predictive assay for treatment failure using functional magnetic resonance imaging and clinical prognostic parameters in cervical cancer. *Cancer* **116**(4), 903–912.

McDonald, D. M., and Choyke, P. L. (2003). Imaging of angiogenesis: From microscope to clinic. *Nat. Med.* **9**(6), 713–725.

Mikhaylova, M., Stasinopoulos, I., Kato, Y., Artemov, D., and Bhujwalla, Z. M. (2009). Imaging of cationic multifunctional liposome-mediated delivery of COX-2 siRNA. *Cancer Gene Ther.* **16**(3), 217–226.

Minowa, T., Kawano, K., Kuribayashi, H., Shiraishi, K., Sugino, T., Hattori, Y., Yokoyama, M., and Maitani, Y. (2009). Increase in tumour permeability following TGF-β type I receptor-inhibitor treatment observed by dynamic contrast-enhanced MRI. *Br. J. Cancer* **101**(11), 1884–1890.

Molema, G. (2005). Design of vascular endothelium-specific drug-targeting strategies for the treatment of cancer. *Acta Biochim. Pol.* **52**(2), 301–310.

Molls, M., and Vaupel, P. (2000). The impact of the tumor microenvironment on experimental and clinical radiation oncology and other therapeutic modalities. *In* "Blood Perfusion and Microenvironment of Human Tumors: Implications for Clinical Radiooncology" (L. W. Brady, H.-P. Heilman, and M. Molls, eds.), pp. 1–4. Springer Verlag, New York.

Monzani, E., and La Porta, C. A. (2008). Targeting cancer stem cells to modulate alternative vascularization mechanisms. *Stem Cell Rev.* **4**(1), 51–56.

Moser, T. L., Stack, M. S., Asplin, I., Enghild, J. J., Hojrup, P., Everitt, L., Hubchak, S., Schnaper, H. W., and Pizzo, S. V. (1999). Angiostatin binds ATP synthase on the surface of human endothelial cells. *Proc. Natl. Acad. Sci. USA* **96**(6), 2811–2816.

Munn, L. L. (2003). Aberrant vascular architecture in tumors and its importance in drug-based therapies. *Drug Discov. Today* **8**(9), 396–403.

Neri, D., and Bicknell, R. (2005). Tumour vascular targeting. *Nat. Rev. Cancer* **5**(6), 436–446.

Neuwelt, E. A., Hamilton, B. E., Varallyay, C. G., Rooney, W. R., Edelman, R. D., Jacobs, P. M., and Watnick, S. G. (2009). Ultrasmall superparamagnetic iron oxides (USPIOs): A future alternative magnetic resonance (MR) contrast agent for patients at risk for nephrogenic systemic fibrosis (NSF)? *Kidney Int.* **75**(5), 465–474.

Nguyen, M., Strubel, N. A., and Bischoff, J. (1993). A role for sialyl Lewis-X/A glycoconjugates in capillary morphogenesis. *Nature* **365**(6443), 267–269.

Ogan, M. D., Schmiedl, U., Mosley, M. E., Grodd, W., Paajanen, H., and Brasch, R. C. (1987). Albumin labeled with Gd-DTPA; an intravascular contrast enhancing agent for magnetic resonance blood pool imaging: Preparation and characterization. *Invest. Radiol.* **22**, 665–671.

Ogawa, S. (1990). Oxygenation-sensitive contrast in MR image of rodent brain at high magnetic fields. *Magn. Reson. Med.* **14**, 68–78.

Oh, P., Li, Y., Yu, J., Durr, E., Krasinska, K. M., Carver, L. A., Testa, J. E., and Schnitzer, J. E. (2004). Subtractive proteomic mapping of the endothelial surface in lung and solid tumours for tissue-specific therapy. *Nature* **429**(6992), 629–635.

Ohno, K. P., Pettigrew, K. D., and Rapoport, S. I. (1979). Local cerebral blood flow in the conscious rat as measured with 14C-antipyrine, 14C-iodoantipyrine and 3H-nicotine. *Stroke* **10**(1), 62–67.

Ostergaard, L., Weisskoff, R. M., Chesler, D. A., Gyldensted, C., and Rosen, B. R. (1996). High resolution measurement of cerebral blood flow using intravascular tracer bolus passages. Part I: Mathematical approach and statistical analysis. *Magn. Reson. Med.* **36**(5), 715–725.

Pasqualini, R., Koivunen, E., Kain, R., Lahdenranta, J., Sakamoto, M., Stryhn, A., Ashmun, R. A., Shapiro, L. H., Arap, W., and Ruoslahti, E. (2000). Aminopeptidase N is a receptor for tumor-homing peptides and a target for inhibiting angiogenesis. *Cancer Res.* **60**(3), 722–727.

Pathak, A. P. (2009). Magnetic resonance susceptibility based perfusion imaging of tumors using iron oxide nanoparticles. *Wiley Interdiscip. Rev. Nanomed. Nanobiotechnol.* **1**(1), 84–97.

Pathak, A. P., Artemov, D., and Bhujwalla, Z. B. (2004). A novel system for determining contrast agent concentration in mouse blood in vivo. *Magn. Reson. Med.* **51**(3), 612–615.

Pathak, A. P., Hochfeld, W. E., Goodman, S. L., and Pepper, M. S. (2008a). Circulating and imaging markers for angiogenesis. *Angiogenesis* **11**(4), 321–335.

Pathak, A. P., Rand, S. D., and Schmainda, K. M. (2003). The effect of brain tumor angiogenesis on the in vivo relationship between the gradient echo relaxation rate change (DR2*) and contrast agent (MION) dose. *J. Magn. Reson. Imaging* **18**(4), 397–403.

Pathak, A. P., Schmainda, K. M., Ward, B. D., Linderman, J. R., Rebro, K. J., and Greene, A. S. (2001). MR-derived cerebral blood volume maps: Issues regarding histological validation and assessment of tumor angiogenesis. *Magn. Reson. Med.* **46**(4), 735–747.

Pathak, A. P., Ward, B. D., and Schmainda, K. M. (2008b). A novel technique for modeling susceptibility-based contrast mechanisms for arbitrary microvascular geometries: The finite perturber method. *Neuroimage* **40**(3), 1130–1143.

Patlak, C. S., Blasberg, R. G., and Fenstermacher, J. D. (1983). Graphical evaluation of blood-to-brain transfer constants from multiple-time uptake data. *J. Cereb. Blood Flow Metab.* **3**, 1–7.

Penet, M. F., Pathak, A. P., Raman, V., Ballesteros, P., Artemov, D., and Bhujwalla, Z. M. (2009). Noninvasive multiparametric imaging of metastasis-permissive microenvironments in a human prostate cancer xenograft. *Cancer Res.* **69**(22), 8822–8829.

Persigehl, T., Bieker, R., Matuszewski, L., Wall, A., Kessler, T., Kooijman, H., Meier, N., Ebert, W., Berdel, W. E., Heindel, W., Mesters, R. M., and Bremer, C. (2007). Antiangiogenic tumor treatment: Early noninvasive monitoring with USPIO-enhanced MR imaging in mice. *Radiology* **244**(2), 449–456.

Provent, P., Benito, M., Hiba, B., Farion, R., Lopez-Larrubia, P., Ballesteros, P., Remy, C., Segebarth, C., Cerdan, S., Coles, J. A., and Garcia-Martin, M. L. (2007). Serial in vivo spectroscopic nuclear magnetic resonance imaging of lactate and extracellular pH in rat gliomas shows redistribution of protons away from sites of glycolysis. *Cancer Res.* **67**(16), 7638–7645.

Raman, V., Artemov, D., Pathak, A. P., Winnard, P. T., Jr., McNutt, S., Yudina, A., Bogdanov, A., Jr., and Bhujwalla, Z. M. (2006). Characterizing vascular parameters in hypoxic regions: A combined magnetic resonance and optical imaging study of a human prostate cancer model. *Cancer Res.* **66**(20), 9929–9936.

Rettig, W. J., Garin-Chesa, P., Healey, J. H., Su, S. L., Jaffe, E. A., and Old, L. J. (1992). Identification of endosialin, a cell surface glycoprotein of vascular endothelial cells in human cancer. *Proc. Natl. Acad. Sci. USA* **89**(22), 10832–10836.

Ribatti, D. (2009). Early evidence of the vascular phase and its importance in tumor growth. *In* "History of Research on Tumor Angiogenesis" (D. Ribatti, ed.), pp. 1–17. Springer, Netherlands.

Roberts, H. C., Roberts, R. C., Brasch, R. T., and Dillon, W. P. (2000). Quantitative measurement of microvascular permeability in human brain tumors achieved using dynamic contrast-enhanced MR imaging: Correlation with histologic grade. *AJNR Am J Neuroradiol* **21**(5), 891–899.

Rosen, B. R., Belliveau, J. W., Buchbinder, B. R., McKinstry, R. C., Porkka, L. M., Kennedy, D. N., Neuder, M. S., Fisel, C. R., Aronen, H. J., Kwong, K. K., Weisskoff, R. M., Cohen, M. S., *et al.* (1991). Contrast agents and cerebral hemodynamics. *Magn. Reson. Med.* **19**, 285–292.

Sato, T. N., Tozawa, Y., Deutsch, U., Wolburg-Buchholz, K., Fujiwara, Y., Gendron-Maguire, M., Gridley, T., Wolburg, H., Risau, W., and Qin, Y. (1995). Distinct roles of the receptor tyrosine kinases Tie-1 and Tie-2 in blood vessel formation. *Nature* **376**(6535), 70–74.

Schlingemann, R. O., Rietveld, F. J., de Waal, R. M., Bradley, N. J., Skene, A. I., Davies, A. J., Greaves, M. F., Denekamp, J., and Ruiter, D. J. (1990). Leukocyte antigen CD34 is expressed by a subset of cultured endothelial cells and on endothelial abluminal microprocesses in the tumor stroma. *Lab. Invest.* **62**(6), 690–696.

Schmainda, K. M., Rand, S. D., Joseph, A. M., Lund, R., Ward, B. D., Pathak, A. P., Ulmer, J. L., Badruddoja, M. A., and Krouwer, H. G. (2004). Characterization of a first-pass gradient-echo spin-echo method to predict brain tumor grade and angiogenesis. *AJNR Am. J. Neuroradiol.* **25**(9), 1524–1532.

Schmieder, A. H., Caruthers, S. D., Zhang, H., Williams, T. A., Robertson, J. D., Wickline, S. A., and Lanza, G. M. (2008). Three-dimensional MR mapping of angiogenesis with alpha5beta1 (alpha nu beta3)-targeted theranostic nanoparticles in the MDA-MB-435 xenograft mouse model. *FASEB J.* **22**(12), 4179–4189.

Schwarzbauer, C., Syha, J., and Haase, A. (1993). Quantification of regional cerebral blood volumes by rapid T_1 mapping. *Magn. Reson. Med.* **29**, 709–712.

Semenza, G. L. (2010). Defining the role of hypoxia-inducible factor 1 in cancer biology and therapeutics. *Oncogene* **29**(5), 625–634.

Senpan, A., Caruthers, S. D., Rhee, I., Mauro, N. A., Pan, D., Hu, G., Scott, M. J., Fuhrhop, R. W., Gaffney, P. J., Wickline, S. A., and Lanza, G. M. (2009). Conquering the dark side: Colloidal iron oxide nanoparticles. *ACS Nano* **3**(12), 3917–3926.

Silva, A. C., Kim, S.-G., and Garwood, M. (2000). Imaging blood flow in brain tumors using arterial spin labeling. *Magn. Reson. Med.* **44**, 169–173.

Sipkins, D. A., Cheresh, D. A., Kazemi, M. R., Nevin, L. M., Bednarski, M. D., and Li, K. C. (1998). Detection of tumor angiogenesis in vivo by alphaVbeta3-targeted magnetic resonance imaging. *Nat. Med.* **4**(5), 623–626.

Sorensen, A. G., Batchelor, T. T., Zhang, W. T., Chen, P. J., Yeo, P., Wang, M., Jennings, D., Wen, P. Y., Lahdenranta, J., Ancukiewicz, M., di Tomaso, E., Duda, D. G., *et al.* (2009). A "vascular normalization index" as potential mechanistic biomarker to predict survival after a single dose of cediranib in recurrent glioblastoma patients. *Cancer Res.* **69**(13), 5296–5300.

Croix, B., St, Rago, C., Velculescu, V., Traverso, G., Romans, K. E., Montgomery, E., Lal, A., Riggins, G. J., Lengauer, C., Vogelstein, B., and Kinzler, K. W. (2000). Genes expressed in human tumor endothelium. *Science* **289**(5482), 1197–1202.

Tarli, L., Balza, E., Viti, F., Borsi, L., Castellani, P., Berndorff, D., Dinkelborg, L., Neri, D., and Zardi, L. (1999). A high-affinity human antibody that targets tumoral blood vessels. *Blood* **94**(1), 192–198.

Thomlinson, R. H., and Gray, L. H. (1955). The histological structure of some human lung cancers and the possible implications for radiotherapy. *Br. J. Cancer* **9**(4), 539–549.

Thompson, H. K. J., Starmer, C. F., Whalen, R. E., and McIntosh, H. D. (1964). Indicator transit time considered as a gamma-variate. *Circulation Res.* **XIV,** June.

Tofts, P. S. (1997). Modeling tracer kinetics in dynamic Gd-DTPA MR imaging. *J. Magn. Reson. Imaging* **7**(1), 91–101.

Tofts, P. S., Brix, G., Evelhoch, J. L., Buckley, D. L., Henderson, E., Larsson, H. B., Knopp, M. V., Lee, T. Y., Parker, G. J., Mayr, N. A., Taylor, J., Port, R. E., et al. (1999). Estimating kinetic parameters from dynamic contrast-enhanced T(1)-weighted MRI of a diffusable tracer: Standardized quantities and symbols. *J. Magn. Reson. Imaging* **10**(3), 223–232.

Torchilin, V. P. (2005). Recent advances with liposomes as pharmaceutical carriers. *Nat. Rev. Drug Discov.* **4**(2), 145–160.

van der Schaft, D. W., Seftor, R. E., Seftor, E. A., Hess, A. R., Gruman, L. M., Kirschmann, D. A., Yokoyama, Y., Griffioen, A. W., and Hendrix, M. J. (2004). Effects of angiogenesis inhibitors on vascular network formation by human endothelial and melanoma cells. *J. Natl. Cancer Inst.* **96**(19), 1473–1477.

Varallyay, C. G., Muldoon, L. L., Gahramanov, S., Wu, Y. J., Goodman, J. A., Li, X., Pike, M. M., and Neuwelt, E. A. (2009). Dynamic MRI using iron oxide nanoparticles to assess early vascular effects of antiangiogenic versus corticosteroid treatment in a glioma model. *J. Cereb. Blood Flow Metab.* **29**(4), 853–860.

Vaupel, P., Kallinowski, F., and Okunieff, P. (1989). Blood flow, oxygen and nutrient supply, and metabolic microenvironment of human tumors: A review. *Cancer Res.* **49**(23), 6449–6465.

Villringer, A., Rosen, B. R., Belliveau, J. W., Ackerman, J. L., Lauffer, R. B., Buxton, R. B., Chao, Y. S., Wedeen, V. J., and Brady, T. J. (1988). Dynamic imaging with lanthanide chelates in normal brain: Contrast due to magnetic susceptibility effects. *Magn. Reson. Med.* **6,** 164–174.

Virchow, R. (1863). Die krankhaften Geschwultste. Hirschwald, Berlin.

Weisskoff, R. M. (1993). Pitfalls in MR measurement of tissue blood flow with intravascular tracers: Which mean transit time. *Magn. Reson. Med.* **29,** 553–559.

Weisskoff, R. M., Zuo, C. S., Boxerman, J. L., and Rosen, B. R. (1994). Microscopic susceptibility variation and transverse relaxation: Theory and experiment. *Magn. Reson. Med.* **31,** 601–610.

Weissleder, R., Cheng, H. C., Marecos, E., Kwong, K., and Bogdanov, A., Jr. (1998). Non-invasive in vivo mapping of tumour vascular and interstitial volume fractions. *Eur. J. Cancer* **34**(9), 1448–1454.

Willett, C. G., Duda, D. G., di Tomaso, E., Boucher, Y., Ancukiewicz, M., Sahani, D. V., Lahdenranta, J., Chung, D. C., Fischman, A. J., Lauwers, G. Y., Shellito, P., Czito, B. G., et al. (2009). Efficacy, safety, and biomarkers of neoadjuvant bevacizumab, radiation therapy, and fluorouracil in rectal cancer: A multidisciplinary phase II study. *J. Clin. Oncol.* **27**(18), 3020–3026.

Yablonskiy, D. A. (1994). Theory of NMR signal behavior in magnetically inhomogeneous tissues: The static dephasing regime. *Magn. Reson. Med.* **32,** 749–763.

Yu, H., Su, M. Y., Wang, Z., and Nalcioglu, O. (2002). A longitudinal study of radiation-induced changes in tumor vasculature by contrast-enhanced magnetic resonance imaging. *Radiat. Res.* **158**(2), 152–158.

Zaharchuk, G. (2007). Theoretical basis of hemodynamic MR imaging techniques to measure cerebral blood volume, cerebral blood flow, and permeability. *AJNR Am. J. Neuroradiol.* **28**(10), 1850–1858.

Zhang, C., Jugold, M., Woenne, E. C., Lammers, T., Morgenstern, B., Mueller, M. M., Zentgraf, H., Bock, M., Eisenhut, M., Semmler, W., and Kiessling, F. (2007). Specific targeting of tumor angiogenesis by RGD-conjugated ultrasmall superparamagnetic iron oxide particles using a clinical 1.5-T magnetic resonance scanner. *Cancer Res.* **67**(4), 1555–1562.

Zhao, D., Jiang, L., Hahn, E. W., and Mason, R. P. (2009). Comparison of 1H blood oxygen level-dependent (BOLD) and 19F MRI to investigate tumor oxygenation. *Magn. Reson. Med.* **62**(2), 357–364.

Zhu, A. X., Holalkere, N. S., Muzikansky, A., Horgan, K., and Sahani, D. V. (2008). Early antiangiogenic activity of bevacizumab evaluated by computed tomography perfusion scan in patients with advanced hepatocellular carcinoma. *Oncologist* **13**(2), 120–125.

Zierler, K. L. (1962). Theoretical basis of indicator-dilution methods for measuring flow and volume. *Circulation Res.* **X,** 393–407, March.

2

An Integrated Approach for the Rational Design of Nanovectors for Biomedical Imaging and Therapy

Biana Godin,* Wouter H. P. Driessen,[†] Bettina Proneth,[†]
Sei-Young Lee,*,[§] Srimeenakshi Srinivasan,*
Rolando Rumbaut,[¶] Wadih Arap,[†,‡] Renata Pasqualini,[†,‡]
Mauro Ferrari,*,[‖],** and Paolo Decuzzi*,[††]

*Department of Nanomedicine and Biomedical Engineering,
The University of Texas Health Science Center, Houston, Texas, USA
[†]David H. Koch Center, The University of Texas M.D. Anderson Cancer
Center, Houston, Texas, USA
[‡]Department of Experimental Diagnostic Imaging, The University of Texas
M.D. Anderson Cancer Center, Houston, Texas, USA
[§]Department of Mechanical Engineering, The University of
Texas at Austin, Austin, Texas, USA
[¶]Children's Nutrition Research Center, Baylor College of Medicine,
Houston, Texas, USA
[‖]Department of Experimental Therapeutic, The University of
Texas M. D. Anderson Cancer Center, Houston, Texas, USA
**Department of Biomedical Engineering, Rice University,
Houston, Texas, USA
[††]Center of Bio-/Nanotechnology and Engineering for Medicine,
University of Magna Graecia, Catanzaro, Italy

Advances in Genetics, Vol. 69
Copyright 2010, Elsevier Inc. All rights reserved.
0065-2660/10 $35.00
DOI: 10.1016/S0065-2660(10)69009-8

ABSTRACT

The use of nanoparticles for the early detection, cure, and imaging of diseases has been proved already to have a colossal potential in different biomedical fields, such as oncology and cardiology. A broad spectrum of nanoparticles are currently under development, exhibiting differences in (i) size, ranging from few tens of nanometers to few microns; (ii) shape, from the classical spherical beads to discoidal, hemispherical, cylindrical, and conical; (iii) surface functionalization, with a wide range of electrostatic charges and biomolecule conjugations. Clearly, the library of nanoparticles generated by combining all possible sizes, shapes, and surface physicochemical properties is enormous. With such a complex scenario, an integrated approach is here proposed and described for the rational design of nanoparticle systems (nanovectors) for the intravascular delivery of therapeutic and imaging contrast agents. The proposed integrated approach combines multi-scale/multiphysics mathematical models with *in vitro* assays and *in vivo* intravital microscopy (IVM) experiments and aims at identifying the optimal combination of size, shape, and surface properties that maximize the nanovectors localization within the diseased microvasculature. © 2010, Elsevier Inc.

I. INTRODUCTION

The use of nanoparticles as carriers for therapeutic and imaging contrast agents is based on the concurrent, expected advantages of homing at the diseased site (as cancer lesions), and the ability to bypass the biological barriers encountered between the point of administration and the target tissue. Oncology is the field of medicine in which there has been remarkable contribution in the field of nanotechnology during the last decade (Heath and Davis, 2008; Nie *et al.*, 2007; Riehemann *et al.*, 2008). Liposomes are the most investigated drug delivery nanoparticles and commercially available since 1996, when liposomal doxorubicin had been granted FDA approval for use against Kaposi's sarcoma. Currently, it is also approved for treatment of metastatic breast cancer and recurrent ovarian cancer.

Since then, a plethora of nanoparticle-based drug delivery systems have been presented and are being developed with different features and multiple functionalities (Ferrari, 2005a,b; Wang *et al*, 2008). These exhibit differences in (i) sizes, ranging from few tens of nanometers (as for dendrimers, gold and iron-oxide nanoparticles) to few hundreds of nanometers (as for polymeric and lipid-based particles) to micron-sized particles; (ii) shapes, from the classical spherical beads to discoidal, hemispherical, cylindrical, and conical; (iii) surface functionalizations, with a broad range of electrostatic charges and biomolecule conjugations. Clearly, the library generated by combining all possible sizes, shapes, and surface physicochemical properties of the nanoparticles currently under development is enormous and this leads naturally to posing the following question: is there any optimal combination that could maximize the accumulation of intravascularly injected nanoparticles at the biological target site (as the cancer lesion) while minimizing their sequestration by the reticuloendothelial system (RES)?

In this chapter, an integrated approach is proposed to tackle such a problem which is based on combining together mathematical models, *in vitro* characterization assays and *in vivo* experiments. The chapter is organized as follows: after the introduction, in Section II, the different types of nanoparticle-based delivery systems so far developed are reviewed chronologically and three different generations are presented as a possible classification method; in Section III, the mathematical models used to predict the behavior of intravascularly injected nanoparticles are presented focusing the attention on their transport within an authentically complex vascular system and specific/nonspecific interactions with cells of the immune systems (RES cells) as well as cells lining the blood vessel walls (endothelial cells); in Section IV, the assays used for characterizing the geometrical and surface physicochemical properties of the nanoparticles are reviewed with an emphasis on those properties that mostly affect the behavior of the nanoparticles *in vivo* as from the mathematical predictions; in Section V, different strategies for targeting the injected nanoparticles to the diseased vasculature are presented, including the use of phage-display peptides and other conventional ligands; in Section VI, an *in vitro* assays for characterizing the dynamics, adhesive, and internalization performances of nanoparticles underflow are described based on the use of parallel plate flow chamber systems; and finally, in Section VII, the powerful technique of intravital video microscopy for monitoring the *in vivo* behavior of the nanoparticles within the diseased vasculature is presented. The cross-interaction within the different components of the proposed integrated approach for the rational design of nanoparticle-based delivery systems is discussed in Section VIII.

II. THREE GENERATIONS OF NANOVECTORS

Nanovectors can be generally organized into three main subclasses or "generations" as schematically shown in Fig. 2.1 (Ferrari, 2005a,b; Harris and Chess, 2003; Sanhai *et al.*, 2008). In the sequel, the term nanovector will be used to identify systems having nanoscale components for the delivery of therapeutic or contrast agent; and the term targeting will be used only when mentioning the specific binding of particles to the site (e.g., due to the presence of mAb on the particle's surface). When describing passive concentration governed by physical laws the terms "homing," "localization," or "direction" will be used.

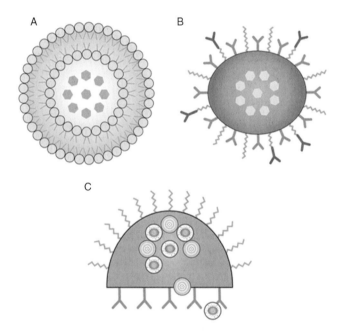

Figure 2.1. (A) First-generation nanovectors, as the currently clinical liposomes, comprise a container (phospholipidic bilayer in yellow) and an active principle (red dots). They localize in the tumor by enhanced permeation and retention (EPR); (B) Second-generation nanovectors further possess the ability for the targeting of their therapeutic action via antibodies and other biomolecules, remote activation, or responsiveness to environment; (C) Third-generation nanovectors such as multistage agents are capable of more complex functions, such a time-controlled deployment of multiple waves of active nanoparticles, deployed across different biological barriers, and with different subcellular targets. (See Color Insert.)

A. The first generation of nanovectors

The first generation of nanovectors (Fig. 2.1A) encompasses a delivery system that localizes into the lesion through passive mechanisms. Liposomes, as the liposomal doxorubicin, home within the tumor tissue mainly through the enhanced permeation and retention (EPR) mechanism. The highly permeable neovasculature of the tumor, characterized by intra- and interendothelium openings large as several hundreds of nanometers, favors the extravasation of nanovectors which simply cross the large vascular fenestrations. These carriers can also be modified with a polyethylene glycol (PEG) or "stealth" layer which prevents their uptake by the RES, thus substantially prolonging the particles' circulation time (Harris and Chess, 2003) and increasing the likelihood of tumor homing (Duncan, 2006; Maeda et al., 2000; Torchilin, 2005). The localization in this case is driven only by the particles' nanodimensions and is not related to any specific recognition of the tumor or neovascular targets. Nonetheless, localization through EPR has been quite successful particularly in changing the pharmacokinetic behavior, bioavailability, and toxicity of the delivered drug. Myocet (non-PEGylated liposomes) and Doxil (PEGylated liposomes) were among the first liposomal systems in clinical use (Drummond et al., 1999; Torchilin, 2005). For the liposomal encapsulated doxorubicin, the elimination half-life for the free drug is only 0.2 h, but this increases to 2.5 and 55 h, respectively, when non-PEGylated and PEGylated liposomal formulations. Moreover, the area under the time–plasma concentration profile (the AUC), which indicates the bioavailability of an agent following its administration, is increased 11- and 200-fold for Myocet and Doxil, respectively, compared to the free drug (Hofheinz et al., 2005). Encapsulation into the liposomal carrier also causes a significant reduction in the most significant adverse side effect of doxorubicin, namely cardiotoxicity, as demonstrated in clinical trials (Drummond et al., 1999; Hofheinz et al., 2005; Parveen and Sahoo, 2006; Torchilin, 2005). Other liposomal drugs which are either currently in use or are being evaluated in clinical trials include non-PEGylated liposomal daunorubicin (DaunoXome) and vincristine (Onco-TCS), PEGylated liposomal cisplatin (SPI-77), and lurtotecan (OSI-211) (Peer et al., 2007; Zhang et al., 2007).

Within the first generation of nanovectors, metal nanoparticles for use in diagnostics, albumin paclitaxel nanoparticles approved for use in metastatic breast cancer, and drug–polymer constructs are also included (Allen, 2002; Gradishar et al., 2005; Peer et al., 2007; Ringsdorf, 1975; Vasey et al., 1999; Zhang et al., 2007).

B. The second generation of nanovectors

The second generation of nanovectors encompasses delivery systems with additional functionality, including surface moieties on the nanovector, providing specific molecular recognition of receptors expressed over the tumor cells or within the tumor vasculature (Fig. 2.1B), and active/triggered release of the payload at the diseased location. The most investigated example of the second-generation nanovectors are antibody-targeted nanoparticles, such as mAb-conjugated liposomes (Adams *et al.*, 2001; Allen, 2002; Banerjee *et al.*, 2001; Brannon-Peppas and Blanchette, 2004; Goren *et al.*, 1996; Juweid *et al.*, 1992; Langer, 1998; Maeda *et al.*, 2000; Saul *et al.*, 2006; Souza *et al.*, 2006; Torchilin, 2007; Yang *et al.*, 2006). A variety of targeting moieties besides antibodies are under extensive investigation worldwide, including ligands, aptamers, small peptides, and phage-display peptide binding to specific target cell-surface markers or surface markers expressed in the disease microenvironment, and these will be further discussed in our chapter (Hajitou *et al.*, 2006a,b; Kang *et al.*, 2008; Sergeeva *et al.*, 2006).

The choice between high or low binding affinity of the ligand for its antigen or receptor remains an issue still to be resolved. Within the tumor, when the binding affinity is high, there is some evidence that the penetration of targeted therapeutics into a cancerous mass is drastically reduced due to the "binding-site barrier." In this case, the targeted therapeutics binds strongly to the first targets encountered, but fails to diffuse further into the tumor. On the other hand, for targets in which most of the cells are readily accessible to the delivery system—for example, tumor vasculature and certain hematological malignancies—a high binding affinity is desirable.

As regarding responsive drug delivery systems, pH- or enzyme-sensitive polymers as well as a diverse group of remotely activated nanovectors have been demonstrated. Among the most interesting examples here are gold nanoshells that are activated by near-infrared light or iron-oxide nanoparticles triggered by oscillating magnetic fields (Duncan, 2003; Hirsch *et al.*, 2003). Linking nanoshells to antibodies that recognize cancer cells enables these novel systems to seek out their cancerous targets prior to applying near-infrared light to heat them up. For example, in a mouse model of prostate cancer, nanoparticles activated with fluoropyrimidine RNA aptamers that recognized the extracellular domain of the PSMA and loaded with docetaxel as a cytostatic drug were used for targeting and destroying cancer cells (Farokhzad *et al.*, 2006a,b). Other techniques used to remotely activate the second-generation particulates include ultrasound and radiofrequency (Douziech-Eyrolles *et al.*, 2007; Monsky *et al.*, 2002; Schroeder *et al.*, 2007). Although the representatives of the second generation have not yet been approved by FDA, there are numerous ongoing clinical trials involving targeted nanovectors, especially in cancer applications.

Compared to their predecessors, the second-generation nanoparticles offered new degrees of sophistication by employing additional complexities such as targeting moieties, remote activation, and environmentally sensitive components. Predominantly, the second generation represents simply a progressive evolution of the first-generation nanovectors. The subtle, yet augmenting, improvements described above do not fully address the primary challenge—or set of challenges—presented in the form of sequential biological barriers that continue to impair the efficacy of delivery systems. This fundamental problem has given rise to a paradigm shift in the design of nanoparticles with the emergence of a third generation of particle that is specifically engineered to avoid biological barriers and to codeliver multiple nanovector payloads with tumor specificity. The ideal injected chemotherapeutic strategy is envisioned to be capable of navigating through the vasculature after intravenous administration, to reach the desired tumor site at full concentration, and to selectively kill cancer cells with a cocktail of agents with minimal harmful side effects.

C. The third generation of nanovectors

Third-generation nanovectors (Fig. 2.1C), such as multistage agents, are capable of more complex functions which enable sequential overcoming of multiple biobarriers. An example is the time-controlled release of multiple payloads of active nanoparticles, negotiating different biological barriers and with different subcellular targets (Sakamoto et al., 2007). This generation of nanovectors represents the first wave of next-generation nanotherapeutics that are specifically equipped to address biological barriers to improve payload delivery at the tumor site. Some of the above-mentioned and most notable challenges include physiological barriers (i.e., the RES, epithelial/endothelial membranes, cellular drug extrusion mechanisms) and biophysical barriers (i.e., interstitial pressure gradients, expression and density of specific tumor receptors, and ionic molecular pumps) (Jain, 1999; Jain and Baxter, 1988). Biobarriers are sequential in nature, and therefore the probability of reaching the therapeutic objective is the product of individual probabilities of overcoming each and every one of them (Ferrari, 2005a,b). By definition, third-generation nanoparticles have the ability to perform a time sequence of functions which involve the cooperative coordination of multiple nanoparticles and/or nanocomponents. This novel generation of nanotherapeutics is exemplified through the employment of multiple nanobased products that synergistically provide distinct functionalities. Here, the nanocomponents will include any engineered or artificially synthesized nanoproducts, such as peptides, oligonucleotides (e.g., thioaptamers, siRNA), and phage with targeting peptides. Naturally existing biological molecules, for example, antibodies, will be excluded from this designation, despite their ability to be synthesized.

Third-generation nanovectors have been developed to address the numerous challenges responsible for reducing the chemotherapeutic efficacy of earlier strategies. For example, surface modification of the exterior of nanoparticles with PEG has proven to be effective in increasing the circulation time within the bloodstream; however, this preservation tactic proves detrimental to the biological recognition and targeting ability of the nanovector (Klibanov et al., 1991). In order to avoid such contradictory approaches of employing incapacitating improvements to therapeutic delivery systems, many research groups are combining multiple nanotechnologies to exploit the additive contributions of the constituent components. One example of third-generation nanoparticles is the biologically active molecular networks known as "nanoshuttles" (Souza *et al.*, 2008). These self-assemblies of gold nanoparticles within a bacteriophage matrix combine the biological targeting capabilities phage-display peptides with hyperthermic response to near-infrared radiation, CT imaging contrast, and surface-enhanced Raman scattering detection of the gold nanoparticles.

The next example of third-generation nanoparticles is the disease-inspired approach of the "nanocell," which are nested nanoparticle constructs that comprise a lipid-based nanoparticle enveloping a polymeric nanoparticle core. In this case, a conventional chemotherapeutic drug (e.g., doxorubicin) is conjugated to a polymer core, and an anti-angiogenic agent (combretastatin) is then trapped within the lipid envelope. When the nanocells are accumulated within the tumor through the EPR effect, the sequential time release of the anti-angiogenic agent, followed by the cytotoxic drug, causes an initial disruption of tumor vascular growth and effectively traps the drug-conjugated nanoparticle core within the tumor to allow an eventual delivery of the cancer cell-killing agent (Sengupta *et al.*, 2005).

The final example of third-generation nanoparticle technology utilizes a multistage approach that addresses many biological barriers experienced by an injectable therapy. Currently, research groups are developing nanoporous silicon microparticles that utilize their unique particle size, shape, and other physical characteristics in concert with active tumor biological targeting moieties to efficiently deliver payloads of nanoparticles to the tumor site, resolving mission-critical issues that must be addressed in a sequential manner when developing drug delivery systems to fight cancer. The multistage drug delivery system is predicated upon a Stage 1 nanoporous silicon microparticle that is specifically designed (through mathematical modeling) to exhibit superior margination and adhesion properties during its negotiation through the systemic blood flow *en route* to the tumor site. Particle characteristics such as size, shape, porosity, and charge can be exquisitely controlled with precise reproducibility through microfabrication techniques. In addition to its favorable physical characteristics, the Stage 1 particle can be surface-treated with such modifications as PEG for RES avoidance and also equipped with biologically active targeting moieties (e.g., aptamers, peptides, phage, antibodies) to enhance the tumor vasculature targeting specificity. This approach decouples the challenges of: (i) transporting therapeutic agents to the tumor-associated vasculature and (ii) delivering therapeutic agents to cancer cells. The Stage 1 particles shoulder the burden of efficiently transporting a nanoparticle payload to the adjacent tumor vasculature within the nanoporous structures of its interior, serving as a cargo for the Stage 2 nanoparticles. Stage 2 particles generically represent any nanodimention construct within the approximate diameter range of 5–100 nm and can be efficiently loaded into the Stage 1 particles with gradual release profile (Tasciotti et al., 2008). The multistage drug delivery system is emblematic of third-generation nanoparticle technology, since the strategy combines numerous nanocomponents to deliver multiple nanovectors to a tumor lesion. The Stage 1 particle is rationally designed to enhance particle margination within blood vessels, and to increase particle/endothelium interaction for enhancing the probability of active tumor targeting and adhesion (Ferrari, 2008). In addition to improved hemodynamic physical properties and active biological targeting by utilizing nanocomponents such as aptamers and phage, the Stage 1 particle can also present with specific surface modifications in order to avoid RES uptake and exhibit degradation rates predetermined by nanopore density. Upon tumor recognition and vascular adhesion, a series of nanoparticle payloads may be released in a sequential order predicated upon Stage 1 particle degradation rates and payload conjugation strategies (e.g., environmentally sensitive crosslinking techniques, pH, temperature, enzymatic triggers). The versatility of this platform nanovector multistage delivery particle allows for a multiplicity of applications. Depending upon the nanoparticle "cocktail" loaded within the Stage 1 particle, this third-generation nanoparticle system can provide for the delivery not only of cytotoxic drugs but also of remotely activated hyperthermic nanoparticles, contrast agents, and future nanoparticle technologies.

III. MATHEMATICAL MODELS FOR PREDICTING THE NANOVECTORS' BEHAVIOR

Inspired by the behavior of circulating blood cells, such as leukocytes and platelets, the dynamics of an intravascularly injected nanovector can be broken down into three main events, extensively described by Decuzzi and Ferrari, 2007: (i) transport and margination dynamics along the vascular network, (ii) firm adhesion to the vascular endothelium, and (iii) control of internalization/translocation across the vascular endothelium. The size, shape, and surface physicochemical properties of nanovectors have been shown to affect, at different extent, each of these three basic events, as described in the sequel. A multiscale, multiphysics mathematical model is required to predict the behavior of intravasculalry injected nanovectors combining *three modules*: (i) a *transport module* for the analysis of nanovector transport within an authentically complex vascular network; (ii) a *margination and adhesion module* for analyzing the near-wall dynamics of a single nanovector, including the margination dynamics and firm adhesion to the vessel walls; and (iii) an *uptake module* for analyzing the possible internalization of nanovectors by RES and endothelial cells.

A. Modeling the transport within the authentic vasculature

The *transport module* analyzes the convective and diffusive transport of nanovectors within an authentically complex vascular network in the presence of blood flow. In the absence of external forces other than the hydrodynamic forces associated with blood flow, the 3D advective–diffusive equation can be employed

$$\frac{\partial C}{\partial t} + \nabla\cdot(VC) = \nabla\cdot(\nabla DC) \tag{2.1}$$

where C is the volume concentration of nanovectors, t is the time, V is the local fluid velocity, D is the molecular diffusion coefficient of nanovectors in a quiescent fluid, and ∇ and $\nabla\cdot$ are, respectively, the gradient and divergence operators. The flow field is derived by solving the momentum and mass conservation equations for a non-Newtonian fluid

$$\rho\frac{DV}{Dt} = \nabla\cdot(\mu\nabla V) - \nabla p \quad \nabla\cdot V = 0 \tag{2.2}$$

where ρ is the density of the fluid, p is the hydrodynamic pressure within the vascular network, and μ is the apparent viscosity for a Casson fluid, given by

$$\mu = [K_c + (\tau_o/\sqrt{(\nabla\cdot V)^2})^{1/2}]^2 \tag{2.3}$$

with K_c the Casson's coefficient of viscosity and τ_o the yield stress of the fluid (Fung, 1993). The Casson model has been widely used for reproducing the rheological properties of blood. Although more complex approaches and rheological laws could be used, the Casson model represents a good compromise between computational efficiency and accuracy of the results (Neofytou, 2004).

The system of Eqs. (2.1)–(2.3) is closed by imposing the proper initial and boundary conditions. For the flow field, a characteristic velocity profile, varying with time, is imposed at the inlet of the vascular network and a reference zero pressure is imposed at the outlet of the vascular network, whereas on the vessel surfaces the classical Starling's law (Jain, 1987) can be imposed

$$v_n = L_p[(p_c - p_i) - \sigma\Delta\pi] \tag{2.4}$$

introducing the wall hydraulic conductivity L_p, the hydrostatic capillary pressure p_c, and interstitial fluid pressure p_i; the osmotic reflection coefficient σ; the oncotic pressure difference between the vessel and the interstitial compartment $\Delta\pi$; and the fluid velocity normal to the vessel wall v_n. The case of an impermeable wall can be readily obtained by imposing $L_p = 0$.

Table 2.1. Hydraulic Permeability of Capillaries in Various Organs

Organ	$L_p \times 10^{-8}$ (μm/(Pa s))	Type of endothelium
Brain	3	Continuous
Skin	100	Continuous
Skeletal muscle	250	Continuous
Lung	340	Continuous
Heart	860	Continuous
Gastrointestinal tract	13,000	Fenestrated
Glomerulus in kidney	15,000	Fenestrated

The model allows to account for the permeability of the vessel walls which varies from organ to organ being very small in the brain vasculature and extraordinary large within the RES organs, as the spleen and the liver, and within the tumor vasculature. Physiologically relevant data for L_p are summarized in Table 2.1.

For the transport equation, the auxiliary conditions are a characteristic concentration profile, varying with time, at the inlet and a reference zero concentration at the outlet of the vascular network. On the permeable vessel surface, a Starling-like boundary condition is imposed, whereas on the remaining portion impermeable boundary conditions including the adhesion K_a and decohesion K_d rates can be introduced to account for the nanovector adhesion and detachment under flow.

The boundary value problem described above allows for the analysis of the transport, adhesion, and extravasation of nanovectors within an authentic complex vascular network. The independent parameters that can be changed and that affect the performance of the nanovector *in vivo* are (i) the geometry of the network, as the vessel diameter, length, tortuosity, and branching type and level; (ii) the hydrodynamic conditions as the mean blood velocity; (iii) the permeability of the vessel walls to fluid and nanovectors, which could vary within the network; (iv) the adhesiveness of the vessel walls to nanovectors. As an example, Fig. 2.2 shows the transport and adhesion of nanovectors within a tortuous vessel with a branch. For oncological applications, a list of physiologically relevant values of the biophysical parameters needed for running the simulations of Fig. 2.2 is listed in Table 2.2.

B. Modeling the margination and adhesion dynamics of nanovectors

Red blood cells (RBCs) tend to preferentially accumulate within the core of the vessels, because of their shape and more importantly deformability under flow, leaving a "cell-free layer" in proximity of the wall (Kim *et al.*, 2007;

Figure 2.2. Concentration C of nanovectors transported along an authentically complex vascular network (top) and wall concentration C_w (bottom) of nanovectors at $t = 3$, 6, and 9 s after a bolus injection within the inlet section (direction from left to right).

Table 2.2. Physiologically Relevant Values for the Parameters in Eqs. (2.1)–(2.5)

	Symbol	Value
Hydraulic conductivity	L_p^{normal}	3×10^{-12} m/(Pa s)
	L_p^{tumor}	3×10^{-11} m/(Pa s)
Osmotic reflection coefficient	σ^{normal}	0.9
	σ^{tumor}	0
Solute reflection coefficient	σ_F	0~0.9
Diffusive permeability	p^{normal}	0
	p^{tumor}	10^{-12}–10^{-18} m/s
Oncotic pressure difference	$\Delta\Pi^{normal}$	2.5 kPa
	$\Delta\Pi^{tumor}$	0
Capillary pressure	P_c^{in}	4 kPa
	P_c^{out}	1 kPa
Interstitial pressure	P_i^{normal}	−0.5 to 0.5 kPa
	P_i^{tumor}	0.5–1.0 kPa
Particle radius	r	100 nm
Fluid dynamic viscosity	μ	3.5×10^{-3} Pa s
Fluid density	ρ	1.05×10^3 kg/m^3
Boltzmann thermal energy	$k_B T$	4.142×10^{-21} J

Sharan and Popel, 2001). The endothelial cell glycocalyx, a network containing glycoproteins and proteoglycans bound to the luminal cell membrane, contributes to the RBC-free layer and is increasingly recognized as a physiologically relevant structure. The cell-free layer has a thickness varying with the vessel lumen and mean blood velocity and ranging from ~0.5 μm in small capillaries (5 μm and more in diameter) to several microns in arterioles (100 μm and more in diameter) (Kim *et al.*, 2007; Sharan and Popel, 2001). Since nanovectors are supposed to inspect the vessel walls seeking for biophysical and biological diversities, as for instance the presence of endothelial fenestrations or an over-expression of specific receptor molecules, they should be rationally designed to accumulate in the cell-free layer. In other words, nanovectors should be rationally designed to marginate under the action of the hydrodynamic forces. Margination is a well-known term in physiology used to describe the lateral drift of leukocytes and platelets toward the endothelial walls, a process that enhances the likelihood of interaction of these circulating cells with the vascular walls. However, whilst leukocyte and platelets margination is an active process requiring an interaction with RBCs and the dilatation of the inflamed vessels and consequent blood flow reduction (Goldsmith and Spain, 1984); particle margination can only be achieved by proper rational design. Within the cell-free layer a linear laminar flow exists with a shear rate S, and under such hydrodynamic conditions it has been shown that spherical nanovectors tend to follow the streamlines without drifting towards the vessel walls (Goldman et al., 1967). Only external gravitational and magnetic forces can deviate the spherical beads from following the streamlines (Decuzzi et al., 2005). Differently, nonspherical particles with sufficient inertia have been shown to drift laterally toward the vessel wall within a linear laminar flow (Fig. 2.3). Clearly, a marginating particle has a larger probability of interacting with the vessel walls, but not necessarily

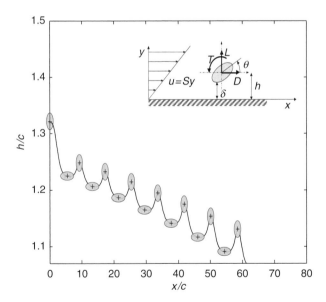

Figure 2.3. The dynamics of an ellipsoidal particle (aspect ratio 0.5) moving in proximity to a vessel wall. The particle is drifting towards the wall at the bottom of the plot (adapted from Lee et al., 2009).

this implies firm adhesion. The propensity for firm adhesion to the vessel walls under flow is related to the probability of adhesion P_a, which is affected by the local hydrodynamic conditions (wall shear rate S); density of the ligand molecules (m_l) distributed over the particle; density of receptor molecules (m_r) expressed on the cell membrane; nonspecific interactions at the cell/particle interface (van der Waals, steric, double layer electrostatic); and particle size and shape (Decuzzi and Ferrari 2006, Decuzzi et al., 2004). Decuzzi and colleagues have demonstrated that oblate spheroidal particles (Decuzzi and Ferrari, 2006) exhibit a much higher strength of adhesion (P_a) compared to spherical particles under the same hydrodynamic and biophysical conditions. In Fig. 2.4, the probability of adhesion is plotted versus the volume of the particle for an oblate spheroid with different aspect ratios γ ranging from unity (spherical particle) to $\gamma = 9$ for which the spheroid degenerates into an almost flat discoidal particle (see bottom of Fig. 2.4). The higher propensity of nonspherical particles to marginate and adhere more strongly to a substrate under flow has been also verified experimentally *in vitro*, as in Fig. 2.5, where the number of particles

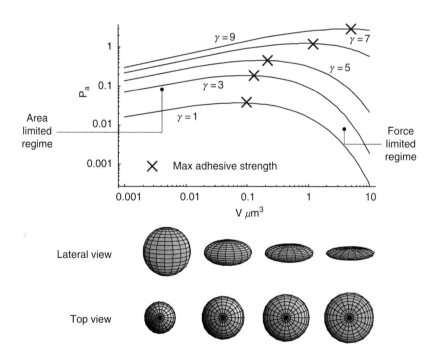

Figure 2.4. The probability of adhesion for a spheroidal particle in a capillary flow (adapted from Decuzzi et al., 2004).

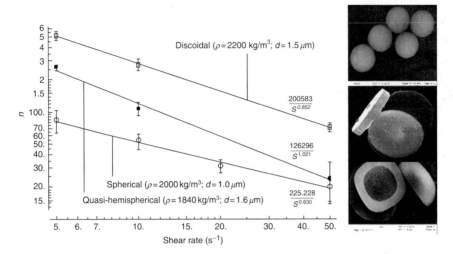

Figure 2.5. Number of marginating nanoparticles as a function of the shear rate and particle shape (adapted from Decuzzi and Ferrari, 2006).

depositing over a layer of fibronectin under a laminar flow in a parallel plate flow chamber has been measured for three different particle geometries (sphere, quasi-hemisphere, and discoidal particles) and different shear rates S.

The margination and adhesion module studies the dynamics of a single nanovector in a linear laminar flow in close proximity of the vessel walls, as within the cell-free layer, accounting for the effect of nanovector size, shape, and specific/nonspecific interactions as well as the contribution of external forces, as gravitational and magnetic. A Lagrangian approach will be used to track particle motion whit the Newtonian governing equations

$$m\frac{\mathrm{d}\boldsymbol{u}}{\mathrm{d}t} = \mathbf{F}, \quad I\frac{\mathrm{d}\boldsymbol{\omega}}{\mathrm{d}t} - I\boldsymbol{\omega} \times \boldsymbol{\omega} = \boldsymbol{T} \tag{2.5}$$

where m and I are the mass and rotational inertia of the particle, respectively; $\boldsymbol{u}$ is the particle velocity vector and $\boldsymbol{\omega}$ is the particle rotational velocity vector; $\boldsymbol{F}$ and $\boldsymbol{T}$ are the force vector and torque exerted over the particle including both surface forces, as the hydrodynamic forces, and external mass forces, as gravitation and magnetic forces. Once the nanovector is in close proximity to the vessel wall (few tens of nanometers), the generalized forces $\boldsymbol{F}$ and $\boldsymbol{T}$ also include the contribution of nonspecific interactions, such as van der Waals, double layer electrostatic, and steric interactions (as for the case of spherical particles in Decuzzi et al., 2005) and of specific interactions, based on the formation/breakage of ligand–receptor bonds as described below.

C. Modeling the cellular internalization of nanovectors

The rate of uptake is affected by the geometry and surface physicochemical properties of the nanovector. This has been clearly shown for spherical particles, where the size (radius) not only affects the uptake rate, but it also affects the uptake mechanisms (Herant *et al.*, 2006; Koval *et al.*, 1998; Rejman *et al.*, 2004). Based on these experimental results, the rate of uptake can be described through a first-order kinetic law where the intracellular concentration $\widetilde{B}(t)$ grows in time following the relationship

$$\frac{\widetilde{B}}{dt} = k_{int}[\chi - \widetilde{B}] \quad k_{int} = \tau_w^{-1} \tag{2.6}$$

where τ_w is the characteristic time for the nanovector to be wrapped by the cell membrane, which can be related to the size, shape, and surface physicochemical properties of the nanovector as explained in the sequel. A mathematical model for receptor-mediated internalization based on an energetic analysis shows that a threshold particle radius exists below which uptake does not occur because it's energetically unfavorable. The same analysis shows that the surface physicochemical properties of the nanovector related to that of the cell membrane can dramatically increase or decrease the uptake rate (Decuzzi and Ferrari, 2007). More recently the effect of particle shape has been also considered by several researchers (Champion and Mitragotri, 2006; Jiang *et al.*, 2008). A mathematical model has been developed to predict the rate of uptake for ellipsoidal particles as a function of their aspect ratio Γ, as shown in Fig. 2.6. Spherical or ovoidal particles can be more rapidly internalized by cells compared to elongated particles. These results clearly emphasize the importance of size, shape, and surface physicochemical properties in controlling the rate of uptake of nanovectors.

D. The adhesion maps for selecting the optimal nanovector

The transport, margination, adhesion, and uptake analyses are then integrated together to generate design maps which recapitulate the performances of nanovectors in terms of transport, specific recognition and adhesion, and uptake as a function of the design parameters and physiological/biophysical conditions. Design maps have been generated in the simpler case of spherical particles as a function of the nonspecific interaction factor F, which accounts for the steric and electrostatic surface interactions between the particle and a cell, and of the ratio β between the number of ligand molecules distributed over the particle surface and the number of receptor molecules expressed over the cell membrane. A representative diagram is shown in Fig. 2.7 for fixed hydrodynamic conditions and ligand–receptor interactions. As a function of F and β, the design maps allow for estimating the propensity of a circulating nanovector to adhere to a specific

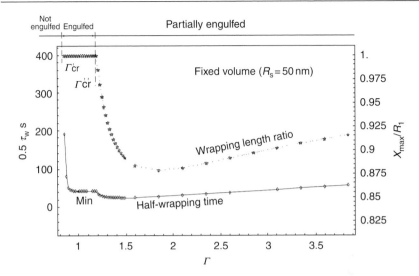

Figure 2.6. Characteristic half-time τ_w for the receptor mediated uptake of ellipsoidal particles with aspect ratio Γ (adapted from Decuzzi and Ferrari, 2007).

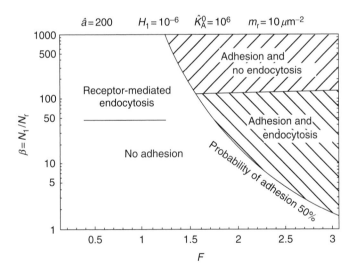

Figure 2.7. Design maps for spherical beads (adapted from Decuzzi and Ferrari, 2008).

vascular target without being internalized by the endothelial cells, or being internalized or even avoiding any adhesion to the endothelial cells (Decuzzi and Ferrari, 2008).

IV. BIOPHYSICAL CHARACTERIZATION OF NANOVECTORS

The geometrical and physicochemical properties of nanovectors are of prime importance for their performance, thus it is vital to characterize drug delivery particles for parameters such as particle size, size distribution, density, surface area, porosity, solubility, surface charge density, purity, surface chemistry, and stability (Donaldson and Borm, 2007). A number of techniques are intrinsically well suited for the characterization of nanoparticles. Methods that are being employed today include among others electron microscopy (transmission and scanning), atomic force microscopy, X-ray diffraction, dynamic light scattering and laser light diffraction, BET, gas pycnometry, and particle sedimentation by gravitation (Donaldson and Borm, 2007). In particular, particle size distribution may be determined by gravitational and centrifugal methods, photon correlation spectroscopy, hydrodynamic chromatography, light scattering methods, electron microscopy methods (Amiji, 2007; Donaldson and Borm, 2007). Shape determination has been possible due to advances in image processing techniques that use shape descriptors, geometric and dynamic, that carry important information to compare shapes of known and unknown particles obtained utilizing high-resolution microscopy (Donaldson and Borm, 2007).

There are advantages and limitations for any of the used techniques and these should be carefully considered. As an example, detailed information of particle sizes, shapes, and agglomeration can be obtained by high-resolution microscopy methods, but these are not robust techniques which are problematic for statistically reliable sampling. In fact, to enable analysis of average diameter and size distribution of the particles, the large number of micrograph sample should be processed by an image analyzing software. Most commonly used high-resolution microscopy techniques are scanning and transmission electron microscopies (SEM and TEM), cryofracture, atomic force microscopy (AFM), and environmental scanning electron microscopy, which also provides information on the elemental analysis of the particle surface (Mohammed *et al.*, 2004; Ruozi *et al.*, 2005). In some cases, various techniques are used to aid in obtaining better resolution images. For example, in SEM analysis, gold sputtering of the sample is frequently utilized to minimize fluctuations of the image thus increasing resolution. In TEM, negative staining is often employed for lipid particles and involves a deposition of an electron opaque metal film (e.g., molybdate) on the sample. The images than appear as bright structures on dark background (a colloidal carbon coated grid; Philippot and Schuber, 1995).

On the other hand, techniques such as dynamic light scattering (DLS), which is based on photon correlation spectroscopy, for analysis of particles size distribution provide us with statistically relevant data. However, in this case, measuring mean particle size, that is, hydrodynamic diameter determined by batch-mode dynamic light scattering in aqueous suspensions, does not give us any information on particles shape and morphology. The DLS method is based on analysis of the time dependence of fluctuations in the intensity of scattered light that results from the Brownian motion of the particulate object in liquid media (Zuidam et al., 2003). The hydrodynamic radius of the object is calculated from the Stocks–Einstein equation, where particle diffusion constant (D) depends on the fluctuations of the diffused light detected by photomultiplier and is in direct relationship with particle dimensions. Particle size distribution can also be determined by gravitational and centrifugal methods, photon correlation spectroscopy, and hydrodynamic chromatography (Barth, 1984; Gotoh et al., 1997; Korgel et al., 1998; Peltonen and Hirvonen, 2008). Particle size is an important physicochemical parameter that affects particles distribution as well as drug loading and release; small particles have a larger surface area, where drugs are usually found and hence lead to fast drug release, while large particles can encapsulate the drug in a core and release it in a slow time dependent manner. Particles' dimensions also have an impact on the ability to cross the endothelial barrier and accumulate in various tissues and organs as will be further discussed in this chapter.

Using the DLS technique, we are able to measure another parameter that plays a pivotal role in biological interactions of the nano- and microparticles, their ζ-potential, which is the potential difference between the dispersion medium and the stationary layer of fluid attached to the dispersed particle. ζ-potential is not measurable directly but it can be calculated using theoretical models and an experimentally determined electrophoretic mobility or dynamic electrophoretic mobility. The electrophoretic mobility and, thus the ζ-potential can be derived from the Doppler shift in frequency of light that is scattered from the particles moving in electric field (Zuidam et al., 2003). The electrostatic properties of the particles depend on the concentration of ions in the solution, the pH of the solution, and the presence of multivalent ions. Other techniques that can be used for electrophoretic mobility measurements are microelectrophoresis and the electroacoustic effects such as colloid vibration current and electric sonic amplitude (Zuidam et al., 2003). ζ-potential measurement brings detailed insight and understanding of the electrokinetic magnitude of the repulsion or attraction between particles in suspension, a degree of agglomeration, particles opsonization, attachment to biological membranes, engulfment by various cells, and finally particles biodistribution, toxicity, and targeting ability.

Particles surface can be modified on their surface with different chemically and biologically active molecules and structures. As an example, hydrophobic particles are quickly opsonized and phagocytosed by the mononuclear phagocyte system, such as the liver, spleen, lungs, and bone marrow, thus, to increase the blood circulation time, particles can be coated with flexible hydrophilic polymers (like PEG), which shield the particles from the RES system (Stolnik et al., 1995). Active targeting to the diseased site with ligands that recognize specific molecular signatures is currently under extensive research. It can be done through attachment of antibodies, ligands, small peptides, or phage-display peptide binding to specific target cell-surface markers or surface markers expressed in the disease microenvironment. There are different techniques available for characterizing the surface conjugated particles and the number of binding sites on the surface of a particle. Among these are fluorimetry, fluorescent and confocal microscopy, scintillation counting, FACS, colorimetry, XPS, AFM, and PCR.

V. VASCULAR TARGETING MOIETIES

Tumor-selective delivery is a much sought-after capability of nanovectors. Target homing strategies can be broadly divided into two categories: passive accumulation and active targeting (Allen *et al.*, 1995; Seymour, 1992; Tyle and Ram, 1990; Neri and Bicknell, 2005). Passive accumulation strategies take advantage of certain unique characteristics of blood vessels in solid tumors; including factors that increase the permeability of vessels (e.g., production of factors mediating vascular permeability such as VEGF (Leung *et al.*, 1989; Senger *et al.*, 1983), nitric oxide (Maeda *et al.*, 1994) and matrix metalloproteases) and the impaired lymphatic clearance of macromolecules and lipids from interstitial tumor tissue leading to enhanced retention. These effects have been well characterized and described over the past years and have been termed the enhanced permeability and retention (EPR) effect (Maeda *et al.*, 2003). Although it is possible to design nanovectors in a way that enhanced tumor-uptake due to the EPR effect is achieved by decorating the particle with polymers such as PEG and HPMA (Andersson *et al.*, 2005), a more rational approach to achieve tumor-selective delivery of nanovectors is active targeting to the vasculature. In this strategy, particles are directed to receptors overexpressed in the tumor microenvironment. As opposed to passive accumulation, this strategy strives to specifically deliver payloads by tumor endothelial cell-surface recognition. Because of the active angiogenesis process and the high vascular density in tumor tissue, tumor endothelial cells provide a large target area, which is readily accessible after systemic administration. Moreover, endothelial cells are genetically stable compared to tumor cells, reducing the likelihood of developing resistance because targets are more stably expressed (Sapra *et al.*, 2005).

The first step in designing vascular-targeted nanovectors is to select the most appropriate target ligand–receptor pair for the downstream application. Ideal targets are expressed on the endothelial cell surface in high numbers and homogenously across the cell population at an accessible location in the vasculature (Ozawa *et al.*, 2008; Sapra and Allen, 2003). The receptor should be inaccessible or expressed in low numbers on nontarget tissues. There have been reports that worked out threshold receptor-densities for achieving improvement of therapeutic efficacy for drug-loaded liposomes targeted to cancer cells. For doxorubicin liposomes targeted with anti-HER2 antibodies, a density of $> 10^5$ Her2-receptors on the cancer cell surface was needed to improve efficacy against an experimental model of metastatic breast cancer (Park et al., 2002). Similar receptor thresholds have been found for GAH- and CD19-targeted cancer-therapies (Hosokawa et al., 2003; Lopes de Menezes et al., 1998). Another consideration is the choice between internalizing versus noninternalizing receptors. Depending on the application of the NanoSystem it could be necessary for the ligand–receptor pair to internalize after the binding event. For delivery of certain therapeutic modalities intracellular delivery is a must (e.g., plasmid DNA or RNAi).

Whole antibodies, antibody-fragments, natural receptor ligands (recombinant or purified), aptamers and low molecular weight ligands such as carbohydrates and peptides, or analogues have been used for active targeting strategies (Allen et al., 2002; Hajitou et al., 2006a,b; Levy-Nissenbaum et al., 2008). The challenge is to identify ligands that have a sufficiently high affinity for their targets. This is even more important when using natural receptor ligands as a targeting moiety, since such a ligand will not only have to recognize and bind to the targeted receptor, but it will also have to compete with the endogenous protein binding to the same receptor-site (Vyas and Sihorkar, 2000). Depending on the downstream application, stability in the circulation is an important parameter also. For drug delivery applications longer circulation times are desirable to maximize therapeutic effects. For imaging-studies using imaging probes with a short half-life short circulation times are of benefit (e.g., positron emission tomography (PET) imaging with 18F as a positron emitter).

Receptors on endothelial cells, which have been targeted in the past, include receptors for angiogenic proteins, adhesion molecules, metabolic receptors, and extracellular matrix components. Numerous studies report the incorporation of targeting moieties into nanoparticles and those of particular interest are summarized below. Emphasis is given to studies which have demonstrated targeting activity in vivo. Among the most widely used vascular targeting agents are peptides containing the arginine-glycine-aspartic acid (RGD) motif directed to the adhesion molecule $\alpha_v\beta_3$-integrin (Pasqualini et al., 1995; Pasqualini et al., 1997). This tripeptide and analogues have been used extensively for numerous applications, including drug- and gene-delivery and molecular imaging (reviewed in Temming et al., 2005). A particularly successful application of this targeting ligand is described by Hood et al. (2002), who used the ligand to target a cationic polymerized lipid-based nanoparticle mixed with a plasmid encoding for the Raf-gene to the tumor vasculature in mice. Pronounced tumor regressions could be achieved after systemic delivery of the gene. Tumor regression could be completely inhibited by preblocking the binding sites with free RGD-peptide, showing specificity (Hood et al., 2002).

Although the notion that a stretch of only three amino acids would serve as such a successful targeting agent was unexpected, there have been more examples of such high-specificity interactions based on short peptide-sequences since. One further example is the NGR-peptide targeting the extracellular matrix component CD13 (Pasqualini et al., 2000). Pastorino et al. (2006) used the peptide to target doxorubicin liposomes to tumor vasculature with antivascular effects in models of lung and ovarian cancer. Interestingly, when this vascular directed treatment was combined with a tumor-targeted therapy (disialoganglioside receptor-targeted immunoliposomes), dramatic improvements in the treatment of an animal model of aggressive metastatic neuroblastoma was found. Other extracellular matrix components which have been targeted are aminopeptidase A (Marchio et al., 2004), ICAM-1, and P-selectin. In one study of interest, ICAM-1-targeted polystyrene particles were used to investigate the influence of size and shape on the particle biodistribution in mice. Carrier geometry was found to influence homing in the vasculature, and the rate of endocytosis and lysosomal transport within ECs. Disks had longer half-lives in circulation and higher targeting specificity in mice, whereas spheres were endocytosed more rapidly. Micron-size carriers had prolonged residency in prelysosomal compartments. Submicron carriers trafficked to lysosomes more readily. Therefore, rational design of carrier geometry will help optimize endothelium-targeted therapeutics and may improve efficacy of enzyme replacement therapy for lysosomal disorders (Muro et al., 2008). Another study used anti-P-selectin-conjugated liposomes containing VEGF for the selective targeting to the infarcted heart. After treatment, changes in cardiac function and vasculature were quantified in a rat model of myocardial ischemia (MI). Specific delivery of VEGF to post-MI tissue resulted in significant increase in fractional shortening and improved systolic function. These functional improvements were accompanied by a 21% increase in the number of anatomical vessels and a 74% increase in the number of perfused vessels in the MI region of treated animals (Scott et al., 2009).

Another class of receptors which have been widely used for vascular targeting are receptors for angiogenic proteins. One obvious example sparked by the marketing of trastuzumab, an antibody directed against Her2, is the targeting of nanosystems to EGF-receptor. Recently, Park and Yoo conjugated an EGF-fragment (11 amino acid peptide) and doxorubicin (DOX) to a bifunctional PEG-polymer. After mixing the conjugate, free DOX and triethylamine "nanoaggregates" spontaneously formed and showed a higher antitumor effect in animals bearing human lung-carcinoma xenografts (Park and Yoo, 2010). In a paper by Diagaradjane et al. (2008), EGF-targeted fluorescent quantum dots were successfully employed for the noninvasive optical imaging of EGFR expression in human colorectal cancer xenografts in mice. Other vascular receptors for angiogenic proteins which can be targeted are VEGF, FGF, TNF, and TGF.

A very active field of vascular-targeting research is focused on directing nanosystems to the blood–brain barrier (BBB). The use of metabolic receptors, such as transferrin-receptor, has proven particularly useful for directing nanoparticles to the BBB and even across it. This approach has been pioneered by the William Pardridge's research-group who have published numerous successes using antibodies and antibody-fragments (see Pardridge, 2008 for a recent review). Using a similar approach, targeting the insulin-receptor, the group has shown global gene-expression in a primate brain after intravenous administration of pegylated immunoliposomes (Zhang *et al.*, 2003).

VI. *IN VITRO* ASSAYS FOR NANOVECTORS' CHARACTERIZATION

Parallel plate flow chambers have been used to characterize the behavior of circulating cells, as leukocytes and platelets, and can be effectively used to characterize the margination, adhesive, and internalization performances of micro/nanoparticles *in vitro*. For instance, Decuzzi and colleagues (Decuzzi *et al.*, 2007; Gentile *et al.*, 2008a; Gentile *et al.*, 2008b) have analyzed the nonspecific adhesion of spherical fluorescent beads (from 50 nm up to 10 μm) to a cellular layer grown on the chamber substrate under flow.

The apparatus, sketched in Fig. 2.8, is composed of (i) parallel plate flow chamber, (ii) up-right fluorescent confocal microscope, (iii) digital camera, (iv) pumping system, (v) connecting tubing, (vi) PC, and (vii) cellular layer. By decorating the nanovector with fluorescent molecules, the transport along the

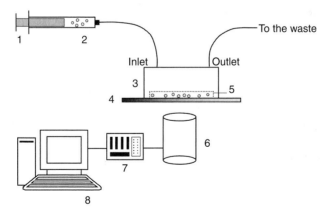

Figure 2.8. The experimental apparatus: (1) syringe (and pump), (2) nanovectors in solution, (3) flow chamber, (4) cell culture dish, (5) gasket, (6) microscope, (7) acquisition system, (8) PC.

chamber and the adhesive dynamics to the bottom cell layer can be analyzed over time. An external pumping system (syringe pump) can be connected to the chip through tubings and by changing the pump infusion rate, the mean fluid velocity within the chamber can be changed accordingly. The chamber is rectangular with a length of 20 mm, a width that can be changed from 2.5 up to 10 mm and a thickness of 100 and 250 μm. However, the flow deck can be customized changing the actual geometry of the chamber. By changing the chamber size and the pump infusion rate, the shear stress at the wall can range from zero to few hundreds of Pa, reproducing different hydrodynamic conditions which are physiologically relevant. Through the digital camera, pictures within the ROI of the moving and stationary particles can be taken at a video rate frequency ($>$30 frames per second) and analyzed off-line. From imaging analysis, (i) the surface density of adherent nanovectors and (ii) the rate of adhesion and decohesion over time can be measured and compared with the mathematical predictions. Images from a typical flow chamber experiment are given in Fig. 2.9, showing, from left to right, fluorescent beads adhering to the cellular layer in

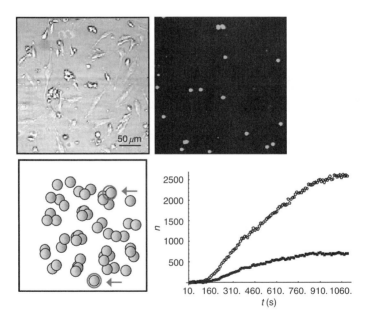

Figure 2.9. From left: Bright-field and fluorescence images of 1 μm fluorescent particles adhering to a subconfluent layer of human umbilical vein endothelial cells (HUVECs); fluorescent image; schematic for image analysis; number of adhering particles over time (Decuzzi *et al.*, 2007). (See Color Insert.)

bright and fluorescent field, imaging analysis, and the final result with the variation of the number of beads adhering to the cell layer within the ROI. The same chamber can be heated up to 37 °C or cooled down to 4 °C to perform internalization experiments over time under flow or in a quiescent medium. Any type of cell growing adherent on a substrate can be used within the flow chamber apparatus.

VII. INTRAVITAL VIDEO MICROSCOPY

Quantitative assessment of microvascular phenomena in real-time, *in vivo*, is feasible using intravital microscopy (IVM). A wide variety of microvascular beds are suitable for IVM, both in tissues *in situ* and surgically exteriorized vascular beds. The experimental techniques involved vary according to the animal species and vascular bed utilized, details are described in representative publications (Harris *et al.*, 2002; Li *et al.*, 2007; Rumbaut *et al.*, 2004). Commonly, these involve anesthesia followed by instrumentation to monitor and maintain the animal's physiologic state and enable administration of exogenous substances (e.g., arterial, venous, tracheal catheters, temperature probes, homeothermic systems, etc.). Figure 2.10 provides examples of three microvascular beds: cremaster muscle, mesentery, and the limbal vessels surrounding the cornea. As evident in the figure, the optical characteristics of thin tissues like the cremaster and mesentery (panels A and B) enable bright-field observations of the interactions between blood cells and microvascular endothelium. In contrast, imaging of thicker tissues such as the limbal vessels of the cornea is typically performed with epifluorescence microscopy (panel C).

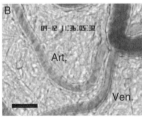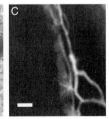

Figure 2.10. Intravital microscopy examples. (A) Bright-field imaging of mouse cremaster showing three rolling leukocytes (arrows) in a convergent venule. (B) Bright-field imaging of rat mesentery with a convergent venule (ven.) and arteriole (art.). (C) Epifluorescence image of limbal microvessels of the mouse cornea, following intravascular injection of a fluorescently labeled macromolecule. Bar = 50 μm.

Quantitative data generated with these techniques include measures of network architecture, vascular diameters, flow patterns, mean blood velocity, and interactions between blood cells and endothelial cells (Li *et al.*, 2006; Rumbaut *et al.*, 2005), among many others. IVM is also used to allow *in situ*, *in vivo*, single microvessel perfusion, for example, to measure microvascular pressures, hydraulic conductivity, or macromolecular permeability (Rumbaut and Huxley, 2002; Rumbaut *et al.*, 2000), as well as for electrophysiologic studies of microvascular endothelial and smooth muscle cells (Welsh *et al.*, 1998).

IVM has also been used to monitor the dynamics of micro- and nano-particles in the microcirculation and their interactions with blood cells and endothelial cells (Ravnic *et al.*, 2007; Smith *et al.*, 2003). These studies involve epifluorescence microscopy, since the spatial resolution of bright-field systems is typically insufficient to resolve particles *in vivo*. Using low-intensity epiillumina-tion (to minimize light/dye-induced phototoxicity and photobleaching) and high-sensitivity imaging, the kinetics of individual fluorescently labeled particles may be resolved reliably, as depicted in Fig. 2.11.

Further, combining this approach with flash epiillumination has been used to estimate fluid velocity profiles in the plasma region near the vessel wall (Smith *et al.*, 2003). These studies reveal the presence of a hydrodynamically significant layer on the surface of endothelial cells, of comparable thickness to those obtained using other IVM-based optical approaches to estimate the thickness of the glycocalyx *in vivo* ($\sim$0.5 μm, Vink and Duling, 1996).

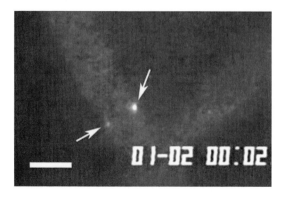

Figure 2.11. Two fluorescently labeled microparticles (1 μm diameter, arrows) in a mouse cremaster arteriole. Bar = 20 μm.

The suitability of specific nanoparticulate systems for IVM depends on various physicochemical properties of the particles (Ravnic et al., 2007), such as size, surface charge, and fluorescence intensity. Further, the signal-to-noise ratio for individual applications is impacted by the sensitivity of the imaging system and a variety of biological parameters, which vary according to the vascular bed. These include tissue autofluorescence and motion artifacts of various frequencies, induced from respiratory excursions, peristalsis, heartbeat, and muscle contraction. Despite these limitations, IVM is a powerful tool to assess the in vivo kinetics of nanoparticles, including their interactions with blood cells and endothelial cells in vivo. These techniques represent a significant component of an integrated approach for the rational design of nanoparticulate systems for biological imaging and therapy.

VIII. CONCLUSIONS

The presented integrated approach combines multiscale/multiphysics mathematical models (Math Tools) with in vitro assays (In vitro Apparatus) and in vivo IVM experiments (IntraVital Microscopy). By doing so, it aims at identifying the optimal combination of size, shape, and surface properties that maximize the nanovectors localization within the diseased microvasculature. The integration among the three fundamental components and the flow of data and information is shown schematically in Fig. 2.12. The geometry of the vascular network, its permeability to plasma, molecules, and nanovectors, and the hemodynamic conditions can be estimated in the authentic vasculature of small animals, as described in Section VII, and the adhesive properties of the nanovectors (association and dissociation coefficients) can be measured using flow chamber systems, as shown in Section VI. These data can be used to accurately define the boundary and initial conditions for the transport, adhesion, and internalization problem presented in Section II. The results of the mathematical models, in terms of number density of nanovectors adhering to the diseased vasculature; and the volume concentration of circulating and extravasated nanovectors can be readily compared with the in vitro and in vivo experimental results to validate the models and further refine the models and parameters used. Such an approach not only helps in identifying the best size, shape, and surface properties combination for a specific nanovector, but it could also dramatically reduce the time and costs for developing and optimize new nanovectors for biomedical applications.

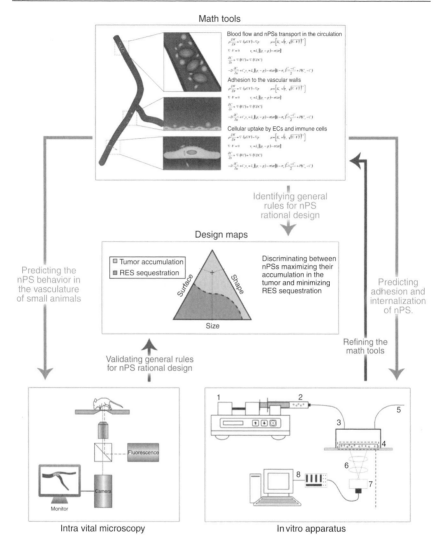

Figure 2.12. The integrated approach and the interaction among the three fundamental components.

Acknowledgments

This work has been partially supported by the Telemedicine and Advanced Technology Research Center (TATRC) of the U.S. Army Medical Research Acquisition Activity (USAMRAA) through the pre-Center Grant entitled "Rational Design of Particulate Systems for the Imaging and Hyperthermia Treatment of an Inflamed Endothelium." and NIH U54CA143837. Matt Landry is gratefully recognized for his assistance in preparing the artwork for this manuscipt.

References

Adams, G. P., Schier, R., McCall, A. M., Simmons, H. H., Horak, E. M., Alpaugh, R. K., and Weiner, L. M. (2001). High affinity restricts the localization and tumor penetration of single-chain fv antibody molecules. *Cancer Res.* **61,** 4750.

Allen, T. M. (2002). Ligand-targeted therapeutics in anticancer therapy. *Nat. Rev. Drug. Discov.* **2,** 750.

Allen, T. M., Ahmad, I., Lopes de Menezes, D. E., and Moase, E. H. (1995). Immunoliposome-mediated targeting of anti-cancer drugs in vivo. *Biochem. Soc. Trans.* **23,** 1073–1079.

Allen, T. M., Sapra, P., Moase, E., Moreira, J., and Iden, D. (2002). Adventures in targeting. *J. Liposome Res.* **12,** 5–12.

Amiji, M. M. (2007). Nanotechnology for cancer therapy. CRC press, Boca Raton, FL, pp. 258–260.

Andersson, L., Davies, J., Duncan, R., Ferruti, P., Ford, J., Kneller, S., Mendichi, R., Pasut, G., Schiavon, O., Summerford, C., Tirk, A., Veronese, F. M., Vincenzi, V., and Wu, G. (2005). Poly (ethylene glycol)-poly(ester-carbonate) block copolymers carrying PEG-peptidyl-doxorubicin pendant side chains: Synthesis and evaluation as anticancer conjugates. *Biomacromolecules* **6,** 914–926.

Banerjee, R. K., van Osdol, W., Bungay, P. M., Sung, C., and Dedrick, R. L. (2001). Finite element model of antibody penetration in a prevascular tumor nodule embedded in normal tissue. *J. Control. Release* **74,** 193.

Barth, H. G. (1984). Modern Methods of Particle Size Analysis. Vol. 73, Wiley-Interscience.

Brannon-Peppas, L., and Blanchette, J. O. (2004). Nanoparticle and targeted systems for cancer therapy. *Adv. Drug Deliv. Rev.* **56,** 1649.

Champion, J. A., and Mitragotri, S. (2006). Role of target geometry in phagocytosis. *Proc. Natl. Acad. Sci. USA* **103,** 4930–4934.

Decuzzi, P., and Ferrari, M. (2006). The adhesive strength of non-spherical particles mediated by specific interactions. *Biomaterials* **27,** 5307–5314.

Decuzzi, P., and Ferrari, M. (2007). The role of specific and non-specific interactions in receptor-mediated endocytosis of nanoparticles. *Biomaterials* **28,** 2915–2922.

Decuzzi, P., and Ferrari, M. (2008). Design maps for nanoparticles targeting the diseased microvasculature. *Biomaterials* **29,** 377–384.

Decuzzi, P., Lee, S., Decuzzi, M., and Ferrari, M. (2004). Adhesion of microfabricated particles on vascular endothelium: A parametric analysis. *Ann. Biomed. Eng.* **32,** 793–802.

Decuzzi, P., Lee, S., Bhushan, B., and Ferrari, M. (2005). A theoretical model for the margination of particles within blood vessels. *Ann. Biomed. Eng.* **33,** 179–190.

Decuzzi, P., Gentile, F., Granaldi, A., Curcio, A., Causa, F., Indolfi, C., Netti, P., and Ferrari, M. (2007). Flow chamber analysis of size effects in the adhesion of spherical particles. *Int. J. Nanomed.* **2,** 689–696.

Decuzzi, P., Pasqualini, R., Arap, W., and Ferrari, M. (2009). Intravascular delivery of particulate systems: Does geometry really matter? *Pharm. Res.* **2009,** 26235–26243.

Diagaradjane, P., Orenstein-Cardona, J. M., Colon-Casasnovas, N. E., Deorukhkar, A., Shentu, S., Kuno, N., Schwartz, D. L., Gelovani, J. G., and Krishnan, S. (2008). Imaging epidermal growth factor receptor expression in vivo: Pharmacokinetic and biodistribution characterization of a bioconjugated quantum dot nanoprobe. *Clin. Cancer Res.* **14,** 731–741.

Donaldson, K., and Borm, P. (2007). Particle Toxicology. CRC Press, Boca Raton, FL, pp. 49–56.

Douziech-Eyrolles, L., Marchais, H., Herve, K., Munnier, E., Souce, M., Linassier, C., Dubois, P., and Chourpa, I. (2007). Nanovectors for anticancer agents based on superparamagnetic iron oxide nanoparticles. *Int. J. Nanomed.* **2,** 541.

Drummond, D. C., Meyer, O., Hong, K., Kirpotin, D. B., and Papahadjopoulos, D. (1999). Optimizing liposomes for delivery of chemotherapeutic agents to solid tumors. *Pharmacol. Rev.* **51,** 691.

Duncan, R. (2003). The dawning era of polymer therapeutics. _Nat. Rev. Drug Discov._ **2**, 347.

Duncan, R. (2006). Polymer conjugates as anticancer nanomedicines. _Nat. Rev. Cancer_ **6**, 688.

Farokhzad, O. C., Cheng, J., Teply, B. A., Sherifi, I., Jon, S., Kantoff, P. W., Richie, J. P., and Langer, R. (2006a). Targeted nanoparticle-aptamer bioconjugates for cancer chemotherapy in vivo. _Proc. Natl. Acad. Sci. USA_ **103**, 6315.

Farokhzad, O. C., Karp, J. M., and Langer, R. (2006b). Nanoparticle-aptamer bioconjugates for cancer targeting. _Exp. Opin. Drug Deliv._ **3**, 311.

Ferrari, M. (2005a). Cancer nanotechnology: Opportunities and challenges. _Nat. Rev. Cancer_ **5**, 161–171.

Ferrari, M. (2005b). Nanovector therapeutics. _Curr. Opin. Chem. Biol._ **9**, 343.

Ferrari, M. (2008). Nanogeometry: Beyond drug delivery. _Nat. Nanotechnol._ **3**, 131.

Fung, Y. C. (1993). Biomechanics: Mechanical Properties of Living Tissues Springer, New York, US.

Gentile, F., Chiappini, C., Fine, D., Bhavane, R. C., Peluccio, M. S., Cheng, M. M., Liu, X., Ferrari, M., and Decuzzi, P. (2008a). The effect of shape on the margination dynamics of non-neutrally buoyant particles in two-dimensional shear flows. _J. Biomech._ **41**, 2312–2318.

Gentile, F., Curcio, A., Indolfi, C., Ferrari, M., and Decuzzi, P. (2008b). The margination propensity of spherical particles for vascular targeting in the microcirculation. _J. Nanobiotechnol._ **6**, 9.

Goldman, A. J. R., Cox, G., and Brenner, H. (1967). Slow viscous motion of a sphere parallel to a plane wall II. Couette flow. _Chem. Eng. Sci._ **22**, 635–660.

Goldsmith, H. L., and Spain, S. (1984). Margination of leukocytes in blood flow through small tubes. _Microvasc. Res._ **27**, 204–222.

Goren, D., Horowitz, A. T., Zalipsky, S., Woodle, M. C., Yarden, Y., and Gabizon, A. (1996). Targeting of stealth liposomes to erbB-2 (Her/2) receptor: In vitro and in vivo studies. _Br. J. Cancer_ **74**, 1749.

Gotoh, K., Masuda, H., and Higashitani, K. (1997). Powder Technology Handbook. Marcel Dekker, New York, US, (pp. 43–55).

Gradishar, W. J., Tjulandin, S., Davidson, N., Shaw, H., Desai, N., Bhar, P., Hawkins, M., and O'Shaughnessy, J. (2005). Phase III trial of nanoparticle albumin-bound paclitaxel compared with polyethylated castor oil-based paclitaxel in women with breast cancer. _J. Clin. Oncol._ **23**, 7794.

Hajitou, A., Pasqualini, R., and Arap, W. (2006a). Vascular targeting: Recent advances and therapeutic perspectives. _Trends Cardiovasc. Med._ **16**, 80–88.

Hajitou, A., Trepel, M., Lilley, C. E., Soghomonyan, S., Alauddin, M. M., Marini, F. C., 3rd., Restel, B. H., Ozawa, M. G., Moya, C. A., Rangel, R., Sun, Y., Zaoui, K., _et al._ (2006b). A hybrid vector for ligand-directed tumor targeting and molecular imaging. _Cell_ **125**, 385.

Harris, J. M., and Chess, R. B. (2003). Effect of pegylation on pharmaceuticals. _Nat. Rev. Drug Discov._ **2**, 214.

Harris, N. R., Whitt, S. P., Zilberberg, J., Alexander, J. S., and Rumbaut, R. E. (2002). Extravascular transport of fluorescently labeled albumins in the rat mesentery. _Microcirculation_ **9**, 177–187.

Heath, J. R., and Davis, M. E. (2008). Nanotechnology and cancer. _Annu. Rev. Med._ **59**, 251.

Herant, M., Heinrich, V., and Dembo, M. (2006). Mechanics of neutrophil phagocytosis: Experiments and quantitative models. _J. Cell Sci._ **119**, 1903–1913.

Hirsch, L. R., Stafford, R. J., Bankson, J. A., Sershen, S. R., Rivera, B., Price, R. E., Hazle, J. D., Halas, N. J., and West, J. L. (2003). Nanoshell-mediated near-infrared thermal therapy of tumors under magnetic resonance guidance. _Proc. Natl. Acad. Sci. USA_ **100**, 13549.

Hofheinz, R. D., Gnad-Vogt, S. U., Beyer, U., and Hochhaus, A. (2005). Liposomal encapsulated anti-cancer drugs. _Anticancer Drugs_ **16**, 691.

Hood, J. D., Bednarski, M., Frausto, R., Guccione, S., Reisfeld, R. A., Xiang, R., and Cheresh, D. A. (2002). Tumor regression by targeted gene delivery to the neovasculature. _Science_ **296**, 2404–2407.

Hosokawa, S., Tagawa, T., Niki, H., Hirakawa, Y., Nohga, K., and Nagaike, K. (2003). Efficacy of immunoliposomes on cancer models in a cell-surface-antigen-density-dependent manner. _Br. J. Cancer_ **89**, 1545–1551.

Jain, R. K. (1987). Transport of molecules across tumor vasculature. *Cancer Metastasis Rev.* **6,** 559–594.

Jain, R. K. (1999). Transport of molecules, particles, and cells in solid tumors. *Annu. Rev. Biomed. Eng.* **1,** 241–263.

Jain, R. K., and Baxter, L. T. (1988). Mechanisms of heterogeneous distribution of monoclonal antibodies and other macromolecules in tumors: Significance of elevated interstitial pressure. *Cancer Res.* **48,** 7022–7032.

Jiang, W., Kim, B. Y., Rutka, J. T., and Chan, W. C. (2008). Nanoparticle-mediated cellular response is size-dependent. *Nat. Nanotechnol.* **3,** 145–150.

Juweid, M., Neumann, R., Paik, C., Perez-Bacete, M. J., Sato, J., van Osdol, W., and Weinstein, J. N. (1992). Micropharmacology of monoclonal antibodies in solid tumors: Direct experimental evidence for a binding site barrier. *Cancer Res.* **52,** 5144.

Kang, J., Lee, M. S., Copland, J. A., III, Luxon, B. A., and Gorenstein, D. G. (2008). Combinatorial selection of a single stranded DNA thioaptamer targeting TGF-beta1 protein. *Bioorg. Med. Chem. Lett.* **18,** 1835.

Kim, S., Kong, R. L., Popel, A. S., Intaglietta, M., and Johnson, P. C. (2007). Temporal and spatial variations of cell-free layer width in arterioles. *Am. J. Physiol. Heart Circ. Physiol.* **293,** H1526–H1535.

Klibanov, A. L., Maruyama, K., Beckerleg, A. M., Torchilin, V. P., and Huang, L. (1991). Activity of amphipathic poly(ethylene glycol) 5000 to prolong the circulation time of liposomes depends on the liposome size and is unfavorable for immunoliposome binding to target. *Biochim. Biophys. Acta* **1062,** 142.

Korgel, B. A., Zanten, J. H., and Monbouquette, H. G. (1998). Vesicle size distributions measured by flow field-flow fractionation coupled with multiangle light scattering. *Biophys. J.* **74,** 3264–3272.

Koval, M., Preiter, K., Adles, C., Stahl, P. D., and Steinberg, T. H. (1998). Size of IgG-opsonized particles determines macrophage response during internalization. *Exp. Cell Res.* **242,** 265–273.

Langer, R. (1998). Drug delivery and targeting. *Nature* **392,** 5.

Lee, S. Y., Ferrari, M., and Decuzzi, P. (2009). Design of bio-mimetic particles with enhanced vascular interaction. *J. Biomech.* **42**(12), 1885–1890.

Leung, D. W., Cachianes, G., Kuang, W. J., Goeddel, D. V., and Ferrara, N. (1989). Vascular endothelial growth factor is a secreted angiogenic mitogen. *Science* **246,** 1306–1309.

Levy-Nissenbaum, E., Radovic-Moreno, A. F., Wang, A. Z., Langer, R., and Farokhzad, O. C. (2008). Nanotechnology and aptamers: Applications in drug delivery. *Trends Biotechnol.* **26,** 442–449.

Li, Z., Rumbaut, R. E., Burns, A. R., and Smith, C. W. (2006). Platelet response to corneal abrasion is necessary for acute inflammation and efficient re-epithelialization. *Invest. Ophthalmol. Vis. Sci.* **47,** 4794–4802.

Li, Z., Burns, A. R., Rumbaut, R. E., and Smith, C. W. (2007). g/d T cells are necessary for platelet and neutrophil accumulation in limbal vessels and efficient epithelial repair after corneal abrasion. *Am. J. Pathol.* **171,** 838–845.

Lopes de Menezes, D. E., Pilarski, L. M., and Allen, T. M. (1998). In vitro and in vivo targeting of immunoliposomal doxorubicin to human B-cell lymphoma. *Cancer Res.* **58,** 3320–3330.

Maeda, H., Noguchi, Y., Sato, K., and Akaike, T. (1994). Enhanced vascular permeability in solid tumor is mediated by nitric oxide and inhibited by both new nitric oxide scavenger and nitric oxide synthase inhibitor. *Jpn. J. Cancer Res.* **85,** 331–334.

Maeda, H., Wu, J., Sawa, T., Matsumura, Y., and Hori, K. (2000). Tumor vascular permeability and the EPR effect in macromolecular therapeutics: A review. *J. Control. Release* **65,** 271.

Maeda, H., Fang, J., Inutsuka, T., and Kitamoto, Y. (2003). Vascular permeability enhancement in solid tumor: Various factors, mechanisms involved and its implications. *Int. Immunopharmacol.* **3,** 319–328.

Marchio, S., Lahdenranta, J., Schlingemann, R. O., Valdembri, D., Wesseling, P., Arap, M. A., Hajitou, A., Ozawa, M. G., Trepel, M., Giordano, R. J., Nanus, D. M., Dijkman, H. B., *et al.* (2004). Aminopeptidase A is a functional target in angiogenic blood vessels. *Cancer cell* **5,** 151–162.

Mohammed, A. R., Weston, N., Coombes, A. G., Fitzgerald, M., and Perrie, Y. (2004). Liposome formulation of poorly water soluble drugs: Optimisation of drug loading and ESEM analysis of stability. *Int. J. Pharm.* **285**(1–2), 23–34.

Monsky, W. L., Kruskal, J. B., Lukyanov, A. N., Girnun, G. D., Ahmed, M., Gazelle, G. S., Huertas, J. C., Stuart, K. E., Torchilin, V. P., and Goldberg, S. N. (2002). Radio-frequency ablation increases intratumoral liposomal doxorubicin accumulation in a rat breast tumor model. *Radiology* **224**, 823.

Muro, S., Garnacho, C., Champion, J. A., Leferovich, J., Gajewski, C., Schuchman, E. H., Mitragotri, S., and Muzykantov, V. R. (2008). Control of endothelial targeting and intracellular delivery of therapeutic enzymes by modulating the size and shape of ICAM-1-targeted carriers. *Mol. Ther.* **16**, 1450–1458.

Neofytou, P. (2004). Comparison of blood rheological models for physiological flow simulation. *Biorheology* **41**(6), 693–714.

Neri, D., and Bicknell, R. (2005). Tumour vascular targeting. *Nat. Rev. Cancer* **5**, 436–446.

Nie, S., Kim, G. J., Xing, Y., and Simons, J. W. (2007). Nanotechnology applications in cancer. *Annu. Rev. Biomed. Eng.* **9**, 257.

Ozawa, M. G., Zurita, A. J., Dias-Neto, E., Nunes, D. N., Sidman, R. L., Gelovani, J. G., Arap, W., and Pasqualini, R. (2008). Beyond receptor expression levels: The relevance of target accessibility in ligand-directed pharmacodelivery systems. *Trends Cardiovasc. Med.* **18**, 126–132.

Pardridge, W. M. (2008). Re-engineering biopharmaceuticals for delivery to brain with molecular Trojan horses. *Bioconjug. Chem.* **19**, 1327–1338.

Park, S., and Yoo, H. S. (2010). In vivo and in vitro anti-cancer activities and enhanced cellular uptakes of EGF fragment decorated doxorubicin nano-aggregates. *Int. J. Pharm.* **383**, 178–185.

Park, J. W., Hong, K., Kirpotin, D. B., Colbern, G., Shalaby, R., Baselga, J., Shao, Y., Nielsen, U. B., Marks, J. D., Moore, D., Papahadjopoulos, D., and Benz, C. C. (2002). Anti-HER2 immunoliposomes: Enhanced efficacy attributable to targeted delivery. *Clin. Cancer Res.* **8**, 1172–1181.

Parveen, S., and Sahoo, S. K. (2006). Nanomedicine: Clinical applications of polyethylene glycol conjugated proteins and drugs. *Clin. Pharmacokinet.* **45**, 965.

Pasqualini, R., Koivunen, E., and Ruoslahti, E. (1995). A peptide isolated from phage display libraries is a structural and functional mimic of an RGD-binding site on integrins. *J. Cell Biol.* **130**, 1189–1196.

Pasqualini, R., Koivunen, E., and Ruoslahti, E. (1997). Alpha v integrins as receptors for tumor targeting by circulating ligands. *Nat. Biotechnol.* **15**, 542–546.

Pasqualini, R., Koivunen, E., Kain, R., Lahdenranta, J., Sakamoto, M., Stryhn, A., Ashmun, R. A., Shapiro, L. H., Arap, W., and Ruoslahti, E. (2000). Aminopeptidase N is a receptor for tumor-homing peptides and a target for inhibiting angiogenesis. *Cancer Res.* **60**, 722–727.

Pastorino, F., Brignole, C., Di Paolo, D., Nico, B., Pezzolo, A., Marimpietri, D., Pagnan, G., Piccardi, F., Cilli, M., Longhi, R., Ribatti, D., Corti, A., Allen, T. M., and Ponzoni, M. (2006). Targeting liposomal chemotherapy via both tumor cell-specific and tumor vasculature-specific ligands potentiates therapeutic efficacy. *Cancer Res.* **66**, 10073–10082.

Peer, D., Karp, J. M., Hong, S. Y., Farokhzad, O., Margalit, R., and Langer, R. (2007). Nanocarriers as an emerging platform for cancer therapy. *Nat. Nanotechnol.* **2**, 751.

Peltonen, L., and Hirvonen, J. (2008). Physicochemical characterisation of nano- and microparticles. *Curr. Nanosci.* **4**(1), 101–107.

Philippot, J. R., and Schuber, F. (1995). Liposomes as Tolls in Basic Research and Industry. CRC press, Boca Raton, FL.

Ravnic, D. J., Zhang, Y. Z., Turhan, A., Tsuda, A., Pratt, J. P., Huss, H. T., and Mentzer, S. J. (2007). Biological and optical properties of fluorescent nanoparticles developed for intravascular imaging. *Microsc. Res. Tech.* **70**, 776–781.

Rejman, J., Oberlem, V., Zuhorn, I. S., and Hoekstra, D. (2004). Size-dependent internalization of particles via the pathways of clathrin- and caveolae-mediated endocytosis. *Biochem. J.* **377**, 159–169.

Riehemann, K., Schneider, S. W., Luger, T. A., Godin, B., Ferrari, M., and Fuchs, H. (2008). Nanomedicine–Developments and perspectives. *Angew. Chem. Int. Ed.* **48**, 872–897.

Ringsdorf, H. (1975). Structure and properties of pharmacologically active polymers. *J. Polym. Sci.* **51**, 135.

Rumbaut, R. E., and Huxley, V. H. (2002). Similar permeability responses to nitric oxide synthase inhibitors of venules from three animal species. *Microvasc. Res.* **64**, 21–31.

Rumbaut, R. E., Wang, J., and Huxley, V. H. (2000). Differential effects of L-NAME on rat venular hydraulic conductivity. *Am. J. Physiol. Heart Circ. Physiol.* **279**, H2017–H2023.

Rumbaut, R. E., Randhawa, J. K., Smith, C. W., and Burns, A. R. (2004). Mouse cremaster venules are predisposed to light/dye-induced thrombosis independent of wall shear rate. CD18. ICAM-1. or P-selectin. *Microcirculation* **11**, 239–247.

Rumbaut, R. E., Slaaf, D. W., and Burns, A. R. (2005). Microvascular thrombosis models in venules and arterioles in vivo. *Microcirculation* **12**, 259–274.

Ruozi, B., Tosi, G., Forni, F., Fresta, M., and Vandelli, M. A. (2005). Atomic force microscopy and photon correlation spectroscopy: Two techniques for rapid characterization of liposomes. *Eur. J. Pharm. Sci.* **25**(1), 81–89.

Sakamoto, J., Annapragada, A., Decuzzi, P., and Ferrari, M. (2007). Antibiological barrier nanovector technology for cancer applications. *Expert Opin. Drug Deliv.* **4**, 359.

Sanhai, W. R., Sakamoto, J. H., Canady, R., and Ferrari, M. (2008). Seven challenges for nanomedicine. *Nat. Nanotechnol.* **3**, 242.

Sapra, P., and Allen, T. M. (2003). Ligand-targeted liposomal anticancer drugs. *Prog. Lipid. Res.* **42**, 439–462.

Sapra, P., Tyagi, P., and Allen, T. M. (2005). Ligand-targeted liposomes for cancer treatment. *Curr. Drug Deliv.* **2**, 369–381.

Saul, J. M., Annapragada, A. V., and Bellamkonda, R. V. (2006). A dual-ligand approach for enhancing targeting selectivity of therapeutic nanocarriers. *J. Control. Release* **114**, 277.

Schroeder, A., Avnir, Y., Weisman, S., Najajreh, Y., Gabizon, A., Talmon, Y., Kost, J., and Barenholz, Y. (2007). Controlling liposomal drug release with low frequency ultrasound: Mechanism and feasibility. *Langmuir* **23**, 4019.

Scott, R. C., Rosano, J. M., Ivanov, Z., Wang, B., Chong, P. L., Issekutz, A. C., Crabbe, D. L., and Kiani, M. F. (2009). Targeting VEGF-encapsulated immunoliposomes to MI heart improves vascularity and cardiac function. *FASEB J* **23**, 3361–3367.

Senger, D. R., Galli, S. J., Dvorak, A. M., Perruzzi, C. A., Harvey, V. S., and Dvorak, H. F. (1983). Tumor cells secrete a vascular permeability factor that promotes accumulation of ascites fluid. *Science* **219**, 983–985.

Sengupta, S., Eavarone, D., Capila, I., Zhao, G., Watson, N., Kiziltepe, T., and Sasisekharan, R. (2005). Temporal targeting of tumour cells and neovasculature with a nanoscale delivery system. *Nature* **436**, 568.

Sergeeva, A., Kolonin, M. G., Molldrem, J. J., Pasqualini, R., and Arap, W. (2006). Display technologies: Application for the discovery of drug and gene delivery agents. *Adv. Drug Deliv. Rev.* **58**, 1622–1654.

Seymour, L. W. (1992). Passive tumor targeting of soluble macromolecules and drug conjugates. *Crit. Rev. Ther. Drug Carrier Syst.* **9**, 135–187.

Sharan, M., and Popel, A. S. (2001). A two-phase model for flow of blood in narrow tubes with increased effective viscosity near the wall. *Biorheology* **38**, 415–428.

Smith, M. L., Long, D. S., Damiano, E. R., and Ley, K. (2003). Near-wall micro-PIV reveals a hydrodynamically relevant endothelial surface layer in venules in vivo. *Biophys. J.* **85**, 637–645.

Souza, G. R., Christianson, D. R., Staquicini, F. I., *et al.* (2006). Networks of gold nanoparticles and bacteriophage as biological sensors and cell-targeting agents. *Proc. Natl. Acad. Sci. USA* **103,** 1215.

Souza, G. R., Yonel-Gumruk, E., Fan, D., Easley, J., Rangel, R., Guzman-Rojas, L., Miller, J. H., Arap, W., and Pasqualini, R. (2008). Bottom-up assembly of hydrogels from bacteriophage and Au nanoparticles: The effect of cis- and trans-acting factors. *PLoS One* **3,** e2242.

Stolnik, S., Illum, L., and Davis, S. S. (1995). Long circulating microparticulate drug carriers. *Adv. Drug Deliv. Rev.* **16,** 195–214.

Tasciotti, E., Liu, X., Bhavane, R., Plant, K., Leonard, A. D., Price, B. K., Cheng, M. M., Decuzzi, P., Tour, J. M., Robertson, F. M., and Ferrari, M. (2008). Mesoporous silicon particles as a multistage delivery system for imaging and therapeutic applications. *Nat. Nanotechnol.* **3,** 151.

Temming, K., Schiffelers, R. M., Molema, G., and Kok, R. J. (2005). RGD-based strategies for selective delivery of therapeutics and imaging agents to the tumour vasculature. *Drug Resist. Updat.* **8,** 381–402.

Torchilin, V. P. (2005). Recent advances with liposomes as pharmaceutical carriers. *Nat. Rev. Drug Discov.* **4,** 145.

Torchilin, V. P. (2007). Targeted pharmaceutical nanocarriers for cancer therapy and imaging. *AAPS J.* **9,** E128.

Tyle, P., and Ram, B. P. (1990). Monoclonal antibodies, immunoconjugates, and liposomes as targeted therapeutic systems. *Targeted Diagn. Ther.* **3,** 3–22.

Vasey, P. A., Kaye, S. B., Morrison, R., Twelves, C., Wilson, P., Duncan, R., Thomson, A. H., Murray, L. S., Hilditch, T. E., and Murray, T. (1999). Phase I clinical and pharmacokinetic study of PK1 [N-(2-hydroxypropyl)methacrylamide copolymer doxorubicin]: First member of a new class of chemotherapeutic agents-drug-polymer conjugates. Cancer Research Campaign Phase I/II Committee. *Clin. Cancer Res.* **5,** 83.

Vink, H., and Duling, B. R. (1996). Identification of distinct luminal domains for macromolecules, erythrocytes, and leukocytes within mammalian capillaries. *Circ. Res.* **79,** 581–589.

Vyas, S. P., and Sihorkar, V. (2000). Endogenous carriers and ligands in non-immunogenic site-specific drug delivery. *Adv. Drug Deliv. Rev.* **43,** 101–164.

Wang, A. Z., Gu, F., Zhang, L., Chan, J. M., Radovic-Moreno, A., Shaikh, M. R., and Farokhzad, O. C. (2008). Biofunctionalized targeted nanoparticles for therapeutic applications. *Expert Opin. Biol. Ther.* **8,** 1063–1070.

Welsh, D. G., Jackson, W. F., and Segal, S. S. (1998). Oxygen induces electromechanical coupling in arteriolar smooth muscle cells: A role for L-type $Ca2+$ channels. *Am. J. Physiol.* **274,** H2018–H2024.

Yang, X., Wang, H., Beasley, D. W., Volk, D. E., Zhao, X., Luxon, B. A., Lomas, L. O., Herzog, N. K., Aronson, J. F., Barrett, A. D., Leary, J. F., and Gorenstein, D. G. (2006). Selection of thioaptamers for diagnostics and therapeutics. *Ann. N. Y. Acad. Sci.* **116,** 1082.

Zhang, Y., Schlachetzki, F., and Pardridge, W. M. (2003). Global non-viral gene transfer to the primate brain following intravenous administration. *Mol. Ther.* **7,** 11–18.

Zhang, L., Gu, F. X., Chan, J. M., Wang, A. Z., Langer, R. S., and Farokhzad, O. C. (2007). Nanoparticles in medicine: Therapeutic applications and developments. *Clin. Pharmacol. Ther.* **83,** 761.

Zuidam, N. J., de Vruch, R., and Crommelin, D. J. A. (2003). *In* "Liposomes" (V. P. Torchilin and V. Weissing, eds.), pp. 31–77. Oxford University Press, New York.

3 Targeted Systemic Gene Therapy and Molecular Imaging of Cancer: Contribution of the Vascular-Targeted AAVP Vector

Amin Hajitou*

*Department of Gene Therapy, Section/ Division of Infectious Diseases, Faculty of Medicine, Imperial College London, Wright-Fleming Institute, St Mary's Campus, Norfolk Place, London, United Kingdom

ABSTRACT

Gene therapy and molecular-genetic imaging have faced a major problem: the lack of an efficient systemic gene delivery vector. Unquestionably, eukaryotic viruses have been the vectors of choice for gene delivery to mammalian cells; however, they have had limited success in systemic gene therapy. This is mainly due to undesired uptake by the liver and reticuloendothelial system, broad tropism for mammalian cells causing toxicity, and their immunogenicity.

Advances in Genetics, Vol. 69
Copyright 2010, Elsevier Inc. All rights reserved.

0065-2660/10 $35.00
DOI: 10.1016/S0065-2660(10)69008-6

On the other hand, prokaryotic viruses such as bacteriophage (phage) have no tropism for mammalian cells, but can be engineered to deliver genes to these cells. However, phage-based vectors have inherently been considered poor vectors for mammalian cells. We have reported a new generation of vascular-targeted systemic hybrid prokaryotic–eukaryotic vectors as chimeras between an adeno-associated virus (AAV) and targeted bacteriophage (termed AAV/phage; AAVP). In this hybrid vector, the targeted bacteriophage serves as a shuttle to deliver the AAV transgene cassette inserted in an intergenomic region of the phage DNA genome. As a proof of concept, we assessed the *in vivo* efficacy of vector in animal models of cancer by displaying on the phage capsid the cyclic Arg-Gly-Asp (RGD-4C) ligand that binds to αv integrin receptors specifically expressed on the angiogenic blood vessels of tumors. The ligand-directed vector was able to specifically deliver imaging and therapeutic transgenes to tumors in mice, rats, and dogs while sparing the normal organs. This chapter reviews some gene transfer strategies and the potential of the vascular-targeted AAVP vector for enhancing the effectiveness of existing systemic gene delivery and genetic-imaging technologies.   © 2010, Elsevier Inc.

I. INTRODUCTION

The major hurdle of gene therapy and genetic imaging has been the inability to deliver vectors at high enough efficiency via a systemic route to the target tissue. Gene therapy has faced problems common to molecular-genetic imaging that uses reporter genes, such as immune response to the vector, large-scale vector production, and specificity. A local delivery of the transgene is necessary as proof-of-principle, but real clinical benefit can only follow systemic delivery in order to reach a large number of cells (i.e., to correct the genetic defect). A successful gene therapy and genetic imaging will depend on the development of efficient systemic vectors, able to provide efficient and long-term expression of the transgene. Such vectors also need to be safe after systemic administration, so that the appropriate genes can be delivered to and expressed in target cells only. It is well known that the main challenges of clinical gene therapy trials are the high cost of vector production and vector targeting. Tissue-targeted gene delivery after systemic administration presents an interesting approach of gene therapy. Enhanced tissue tropism of gene therapy vectors will lead to lower viral load required for therapeutic and detection levels of expression. One approach for targeting is to use promoters that are active only in the target cell (transcriptional targeting; Sadeghi and Hitt, 2005). Although this strategy can reduce or even eliminate potential toxic side effects of the transgene, it does not address the need to avoid those that result from the mislocalization of vector particles.

Furthermore, transcriptional targeting alone is not sufficient to ensure gene expression in the target cell. Thus, further optimization of vectors, especially ligand targeting of vectors, may improve clinical efficiency of gene therapy and molecular-genetic imaging approaches. Ligand targeting of vectors to the angiogenic blood vessels is an attractive approach for targeting. Indeed, in solid tumors, the blood vessels are readily accessible to the circulating ligands and express specific markers that are absent or barely detectable on the normal blood vessels (Hajitou et al., 2006a). Moreover, it is estimated that up of 100 tumor cells are sustained by a single endothelial cell (Folkman, 1997). Finally, vascular cells are genetically more stable and unlikely to acquire resistance to therapy, although this notion has been challenged (Hajitou et al., 2006a; Hida and Klagsburn, 2005).

Many animal viruses have potential for ligand-targeted gene therapy, but require elimination of native tropism for mammalian cells. Commonly used approaches in targeted viruses rely on ablating the native tropism of the viral vectors, retargeting them to alternative receptors, or both. A major drawback of these approaches has been the reduced efficacy resulting from entry via a nonnatural receptor.

The prokaryotic virus bacteriophage has been proposed as a safe vector for targeted systemic delivery of transgenes as it has no intrinsic tropism for mammalian cells and can mediate modest gene expression in these cells after genetic manipulation (Larocca et al., 1998). However, bacteriophage have inherently been considered poor gene delivery vehicles for mammalian cells because they have evolved to infect bacteria only and have no optimized strategies to deliver genes to mammalian cells. We therefore hypothesized that combining the favorable biological attributes of eukaryotic and prokaryotic viruses may facilitate gene therapy targeting applications. We have reported a new generation of targeted hybrid prokaryotic–eukaryotic vectors (Hajitou et al., 2006b) which are chimeras between two single-stranded DNA viruses: (i) an adeno-associated virus (AAV) and (ii) a targeted M13 bacteriophage (termed AAV/phage; AAVP). In our prototype, the AAV transgene cassette is inserted into an intergenomic region of the phage genome and is packaged with the phage DNA into the bacteriophage capsid (Hajitou et al., 2006b). As a proof of concept, we assessed the in vivo efficacy of the vector after systemic delivery in animal models of cancer by displaying on the bacteriophage capsid the cyclic RGD-4C ligand to target αv integrins that are specifically expressed on the tumor blood vessels while absent or barely detectable on the normal blood vessels (Arap et al., 1998; Brooks et al., 1994; Dickerson et al., 2004; Ellerby et al., 1999; Hajitou et al., 2006a,b; Hood et al., 2002). We showed that the vascular-targeted vector was able to systemically and specifically deliver imaging and therapeutic genes to tumors.

II. TARGETED GENE DELIVERY

Systemic gene therapy of cancer has faced a major problem common to other genetic diseases, namely the lack of an efficient systemic vector to produce sustained transgene expression efficiently with limited toxicity. There are many different methods for the delivery of therapeutic genes to cancer cells. Although they can be delivered using "naked DNA" molecules (i.e., a plasmid containing the gene of interest; Bertoni et al., 2006; Meykadeh et al., 2005), usually a vector is needed. Nonviral delivery has entered clinical experimentation, despite the overall minor efficacy.

Viral gene delivery is an attractive strategy for systemic delivery of therapeutic or cytotoxic transgenes. Most progress in viral gene therapy has involved adenovirus (Ad), AAV, and lentivirus (Bouard et al., 2009). Ads have been particularly exploited for cancer gene therapy. The adenoviral genome can be modified in such a way that cancer specificity can be achieved at different stages through the adenoviral life cycle (Green and Seymour, 2002). Some altered Ads have specific deletions in their genome to render them replication incompetent, whereby infection is abortive and no amplification and spread of the adenoviral agent occurs. Adenoviral vectors (Douglas, 2007) raised much hope in the past but generate a great immunogenicity. Another type of vector is based on oncolytic viruses. These vectors are engineered to replicate selectively in cancer cells. The amplified viral progeny can then spread through the tumor and theoretically destroy all cancer cells by virus-mediated cell lysis. In practice, these vectors can also carry therapeutic transgenes if the cancer cell lysis is not efficient enough (Oosterhoff et al., 2005; Kanerva et al., 2005). AAVs have become a very popular gene delivery system, based on their efficiency of transduction (Gregorevic et al., 2004; Wang et al., 2005). AAV vectors exhibit extremely stable expression for the lifetime of a mouse and over several years in larger animal models. Recombinant AAV vectors have been used in a variety of gene therapy applications due to their ability to establish long-term gene expression in vivo (Duan et al., 1998). AAV (a member of the parvovirus family) contains 4.7 kb single-stranded DNA encoding rep and cap open reading frames (ORFs) flanked by two inverted terminal repeats (ITRs, Büning et al., 2008). Rep and cap are replaced by therapeutic transgenes in the recombinant virus leaving ITR cis elements which serve as an origin of replication (Samulski et al., 1987). Upon entry into the cell, the single-stranded AAV DNA genome is converted into the transcriptionally active double-stranded DNA (Duan et al., 1998). A severe disadvantage of AAV vectors for gene therapy is their limited cloning capacity; they cannot accommodate large-size cDNA. Moreover, AAV injection caused an immune reaction against the viral proteins, although recent experiments in dogs have indicated a strategy to overcome this immune response (Wang et al., 2007). Also, the very large viral load required to achieve widespread AAV transduction of tissues *in vivo* using tail vein delivery in mice would challenge current vector production methods, clearly limiting clinical application. Finally, the wide tropism of AAV for mammalian cells leads to transduction of unwanted tissues after intravenous administration. Thus, only regional vascular delivery approach will potentially: (i) facilitate safe passage of the virus for efficient transduction, (ii) allow the use of viral doses to accommodate current limitations imposed by vector production methods, and at the same time (iii) achieve a clinically meaningful outcome.

Targeting of viral vectors for systemic delivery allows for ease of application with body wide access, especially necessary in reaching distant metastatic cancers, and large numbers of somatic cells for correction of monogenic defects. Transcriptional targeting of these vectors has been attempted using tissue-specific promoters (Sadeghi and Hitt, 2005). However, this strategy has two major limitations; first, high titres of virus need to be injected in order to achieve good transgene expression in appropriate cells, and second, immunogenic effects arise from mislocalization of vectors to normal tissues.

A. Ligand targeting of animal viruses: feasibility and challenges

Another targeting strategy of animal viruses involves the manipulation of ligands on viral outer coat proteins to mediate binding to the cells of interest. This often requires the ablation of a virus' natural tropism for mammalian cells to facilitate retargeting to specific tissue types (Waehler *et al.*, 2007). For viral vectors, the ideal approach is to genetically engineer new ligands into the capsid proteins of the virus to generate a single agent to mediate therapy. A number of such ligands have been identified by phage display technology, an *in vivo* screening method in which peptides homing to specific vascular beds are selected after intravenous administration of a phage display random peptide library (Hajitou *et al.*, 2006a). However, addition of such targeting ligands to eukaryotic viruses can alter the structure of the viral capsid and diminish targeting properties of the peptides themselves (Waehler *et al.*, 2007). Indeed, insertion of an exogenous ligand from one structural context into the different structural context of a capsid protein can ablate the function of the ligand or disrupt viral assembly and function. These "context" problems are fundamental, since an ideal candidate peptide ligand may be identified but cannot be applied because the ligand destroys the vector or the vector destroys the ligand. This translation problem stems in part from the fact that peptides isolated from phage libraries are selected in the protein structural context of the pIII capsid protein of the filamentous M13 phage and are then translated into different protein structures of a viral capsid protein. Gosh and Barry (2005) have addressed this context problem for adenoviral vectors, by engineering viral "context-specific" phage libraries by introducing the H and I sheets of the Ad knob domain on to the pIII protein of filamentous bacteriophage. A 12-amino acid (12-mer) random peptide library was constructed by insertion between the H and I sheets in the normal position of the HI loop. Selection of this HI loop context-specific peptide library against C2C12 myoblasts generated a peptide, which when translated back into the knob domain of an adenoviral Ad5 vector, resulted in a vector that was functional, mediating improved muscle cell transduction compared to wild-type Ad5.

B. Phage-based vectors and the hybrid AAVP

Another approach has been to use targeted bacteriophage directly for gene delivery to mammalian cells without the need to transfer peptide from phage display to an eukaryotic vector. Bacteriophage, which is a prokaryotic virus, has no tropism for mammalian cells, but can be engineered to deliver genes to such cells. They are safe and can be targeted by a ligand displayed on their capsid to a specific mammalian receptor after systemic administration. Receptor-mediated internalization of bacteriophage by mammalian cells has been demonstrated by displaying on the phage capsid ligands such as epidermal growth factor (EGF), transferrin, integrin binding ligands, and fibroblast growth factor (FGF2; Larocca et al., 1998). In theory, phage-based vectors have some potential advantages over animal viruses for mammalian cell targeted delivery of transgenes. First, there are no known natural receptors for phage on mammalian cells. Moreover, bacteriophage has been used for antibiotic therapy during the preantibiotic era and is safe even in children after systemic administration (Barrow and Soothill, 1997). In 2006, the U.S. Food and Drug Administration (FDA) announced that it had approved the use of a bacteriophage preparation to be used on ready-to-eat (RTE) meat and poultry products as an antimicrobial agent against *Listeria monocytogenes* (Peek and Reddy, 2006). Finally, unlike mammalian viruses, phage do not require further context modification of their capsid since the targeting peptides are actually selected and isolated directly for targeting specific cell-surface receptors, after screenings of a phage display peptide library. Despite these potential advantages phage-based vectors are considered to be poor vectors as bacteriophage has evolved to infect bacteria only and have no optimized strategies to efficiently express transgenes upon entry into eukaryotic cells. In order to overcome this limitation, we have produced a new generation of hybrid prokaryotic–eukaryotic viral vectors as a chimera between AAV and the filamentous M13 bacteriophage (Hajitou et al., 2006b) and named AAVP (AAV/phage). In this hybrid vector, devoid of any AAV capsid, the targeted bacteriophage particle directs the systemic delivery of the AAV transgene cassette incorporated within the phage genome (Hajitou et al., 2006b, 2007, 2008; Trepel et al., 2009). This vector showed superior gene delivery compared to a regular phage vector with long-term transgene expression *in vivo* after systemic delivery. In the reported studies, we systemically targeted AAVP displaying the cyclic RGD-4C peptide that targets αv integrins overexpressed in the angiogenic blood vessels of tumors (Arap et al., 1998; Brooks et al., 1994; Dickerson et al., 2004; Ellerby et al., 1999; Hood et al., 2002) to specifically deliver genes to the tumor site in mice, rats, and dogs (Hajitou et al., 2006b, 2007, 2008; Paoloni et al., 2009; Tandle et al., 2009; Trepel et al., 2009). This system has potential biological and clinical applications including targeted gene transfer and molecular-genetic imaging. AAVP is a receptor-mediated vector that transduces mammalian cells by binding of the ligand displayed on the phage capsid to the specific receptor. Therefore, in vivo phage display technology can be applied to isolate and identify phage clones able to home and target the tissue of interest after intravenous administration. Indeed, in vivo screenings of phage display peptide library have been used to identify peptide-directed phage clones that home and localize in tumor and normal tissues after intravenous administration (Hajitou et al., 2006a). The selected phage clone displaying a specific peptide can be directly used to generate a tissue-targeted AAVP vector for systemic gene therapy and molecular-genetic imaging.

C. *In vivo* phage display technology and generation of systemic tissue-targeted AAVP vectors

Phage display technology presents a rapid means by which proteins and peptides that bind specifically to predefined molecular targets can be selected and isolated from complex combinatorial peptide libraries. Phage display libraries consist of polypeptides expressed within the coat proteins of filamentous bacteriophage. In the minor pIII coat protein, as many as 10^9 unique peptide sequences can be displayed on the surface of the phage particles (Hajitou et al., 2006a; Pasqualini and Ruoslahti, 1996).

To identify probes that home selectively to various receptors of normal and tumor vasculatures, we and others have used an in vivo phage display technology to isolate peptides that bind selectively to target receptors that are expressed only on certain blood vessels. Both tissue-specific and angiogenesis-related (Hajitou et al., 2006a) vascular ligand receptor pairs have been identified through this technology. In this method, phage libraries are directly administered to mice through the tail vein, and tissues are collected and examined for phage bound to tissue-specific vascular cell markers. In vivo panning has the advantage that the isolated phage displayed peptides home selectively to "intact" targets of interest in vivo; moreover, these ligand peptides may be useful for the functional analysis of the corresponding receptors. By applying in vivo selection of peptides from phage display peptide libraries, several peptides have been identified for selectively targeting the prostate, kidney, skin, pancreas, retina, intestine, uterus, and adrenal gland in mice (Hajitou et al., 2006a; Trepel et al., 2002).

Peptide-directed bacteriophage obtained from *in vivo* phage display technology does not need further capsid modifications and can be used themselves as targeting vectors of gene therapy. Moreover, the bacteriophage is selected for homing and internalizing into the target tissue after systemic delivery. Thus, this approach provides a straightforward way in identifying vascular-accessible vectors. In contrast to protein array systems, it is possible to select binding peptides even if the ligand–receptor interaction is mediated by conformational (rather than linear) epitopes. In summary, ligand-directed screening of combinatorial libraries can be used for the selection of functionally relevant ligand-directed phage that can be developed for targeting gene therapy to molecular targets in the tissue. The ligand-directed bacteriophage selected and identified by the *in vivo* phage display screening can serve to construct and produce a tissue-targeted AAVP vector. Targeted AAVP particles are made as previously described in our detailed protocol (Hajitou *et al.*, 2007). The targeted bacteriophage selected is used to produce a phage-MCS (multicloning site)-based M13 phage that has a MCS inserted into an intergenomic region of its genome. The phage-MCS is then modified to generate the corresponding targeted AAV/phage vector by inserting a recombinant rAAV genome into the MCS of the phage-MCS. This strategy also serves to construct nontargeted control vectors (without ligands) or displaying mutant/scrambled versions of the ligands. In the targeted AAVP prototype vector used to date, the transgene cassette is under the control of a *Cytomegalovirus* (CMV) promoter and flanked by ITRs from AAV serotype 2, and inserted into the intergenomic region of an M13 bacteriophage clone displaying a cyclic RGD-4C ligand on the minor pIII coat protein (Hajitou *et al.*, 2007). Targeted and control vector particles are then amplified, isolated, and purified from the culture supernatant of host bacteria (*Escherichia coli*) as reported (Hajitou *et al.*, 2007). Next, vector particles in suspension are sterile-filtered through 0.45-μm filters, then titrated by infection of host bacteria for colony counting on Luria–Bertani (LB) agar plates under a double antibiotic selection and expressed as bacterial transducing units (TU).

The use of AAV ITRs in the RGD-4C AAVP resulted in improved transduction efficiency by this vector over conventional phage-based vectors and was associated with an improved fate of the delivered transgene, through maintenance of the entire mammalian transgene cassette, better persistence of episomal DNA, and formation of concatamers of the transgene cassette (Hajitou *et al.*, 2006b). AAVP-mediated expression of reporter transgenes was observed in mammalian cells treated with targeted AAVP only, while nontargeted AAVP was unable to bind nor transduce mammalian cells *in vitro* and *in vivo* (Hajitou *et al.*, 2006b, 2008; Tandle *et al.*, 2009).

D. Vascular-targeted systemic gene therapy by RGD-4C AAVP

The progress in clinical cancer gene therapy has been hampered by the poor efficiency of gene transfer that potentially limits the number of vector-transduced tumor cells (Trepel et al., 2009) and thus prevents effective systemic cancer gene therapy. Given the estimates that up of 100 tumor cells are sustained by a single endothelial cell (Folkman, 1997), vascular gene targeting might minimize or overcome this problem. Indeed, a small number of transduced cells that are accessible to the circulation could in theory mediate a much more pronounced effect that is relatively independent of gene transfer efficiency. Vascular targeting of gene delivery to the tumor tissue presents an interesting approach of "theragnostics," that is, both gene therapy and genetic imaging combined into one vector system. In solid tumors, the vasculature is particularly an attractive target for gene therapy because vascular cells are readily accessible through the systemic circulation and express several surface markers that are absent or barely detectable in normal blood vessels (Hajitou et al., 2006a). The use of in vivo phage display technology has significantly contributed to the identification of such targets and their corresponding peptides that home to the angiogenic vasculature of tumors (Table 3.1). These motifs include the sequence RGD-4C that targets and binds to αv integrin receptors. The tumor homing is possible because αv integrins play an important role in angiogenesis: the $\alpha v \beta 3$ and $\alpha v \beta 5$ integrins are absent or expressed at low levels in normal endothelial cells but are induced in the angiogenic vasculature of tumors (Arap et al., 1998; Brooks et al., 1994; Dickerson et al., 2004; Ellerby et al., 1999; Hajitou et al., 2006b; Hood et al., 2002). Aminopeptidase N/CD13 has been identified as an angiogenic receptor for the NGR motif (Pasqualini et al., 2000) and

Table 3.1. Cell Surface Receptors Expressed on Angiogenic Vascular Cells of Cancer

Receptor	Function	Vascular localization
Integrin $\alpha v \beta 3$	Cell adhesion	EC
Integrin $\alpha v \beta 5$	Cell adhesion	EC
Aminopeptidase N/CD13	Protease	EC, pericytes
Aminopeptidase A	Protease	Pericytes
MMP-2/MMP-9	Proteases	EC, pericytes
NG2/HMWMAA	Proteoglycan	EC, pericytes
IL 11-R	Cytokine receptor	EC
VEGFR	Growth factor receptor	EC
PDGFR	Growth factor receptor	Pericytes
HSP90	Heat shock	EC

EC, endothelial cells.

aminopeptidase A as an angiogenic receptor for the CPRECESIC motif (Marchiò et al., 2004). Some of the targets turned out to be matrix metalloproteinase MMP-2/MMP-9 (which also act as receptors in tumor blood vessels) are specifically expressed in angiogenic endothelial cells and pericytes of both human and murine tissue origin. Moreover, targets such as the proteoglycan (NG2) have been identified by phage display technology (Hajitou et al., 2006a). The use of in vivo phage display approach in humans led to the identification of IL-11 receptor (IL-11Rα) as a potential target receptor for intervention in human prostate cancer (Arap et al., 2002). To date several angiogenic receptors of tumor vasculature have been identified and used for vascular targeting of cancer (Hajitou et al., 2006a). These include the heat-shock protein 90 kDa (HSP90), vascular endothelial growth factor receptors (VEGFR), and platelet derived growth factor receptors (PDGFR). Markers in the angiogenic neovasculature are either expressed at very low levels or not at all in nonproliferating endothelial cells. Interestingly, many of these tumor vascular markers are proteases (somewhat intuitive, given that malignant tumors are invasive); some of the markers also serve as receptors for animal viruses. It is tempting to speculate that bacteriophage (i.e., prokaryotic viruses) may use the same cellular receptors as do eukaryotic viruses. In fact, the structure of the phage capsid protein provides good evidence that bacteriophage share ancestry with animal viruses (Hendrix, 1999).

Therapeutic efficiency of the vascular-targeted RGD-4C AAVP vector was first evaluated using vector carrying the herpes simplex virus type I thymidine kinase (HSVtk) gene in combination with the prodrug ganciclovir (GCV). This suicide gene therapy has the advantage that not all the cells need to be transduced in order to observe a significant tumor regression. Indeed, the converted cytotoxic drug and/or toxic metabolites can spread from transduced cells to nontransduced cells. This "bystander effect" may overcome the low tumor transduction and the need for all cells to be transduced in order to achieve therapy. Recently, we showed that although tumor cells are not transduced by HSVtk, they can be killed in vitro and in vivo by HSVtk-transduced endothelial cells, via a heterotypic endothelial cell-mediated bystander effect through gap junctional intercellular communication between endothelial and tumor cells (Trepel et al., 2009).

Efficacy of the RGD-4C AAVP-HSVtk vector was tested in preclinical models of cancer from different species and histologic origins with tumors established in immunosuppressed or immunocompetent mice. Marked suppression of the growth of tumors derived from Kaposi's sarcoma KS1767 cells was observed in immunodeficient nude mice given a single intravenous administration of RGD-4C AAVP-HSVtk as compared with mice treated with vehicle or with nontargeted AAVP (Hajitou et al., 2006b). Similar tumor growth suppressive effects were observed in UC3-derived bladder carcinomas and DU145-derived prostate carcinomas in nude mice, regardless of the size of the xenograft being treated (Hajitou *et al.*, 2006b). As another test, efficacy of RGD-4C AAVP-*HSVtk* was analyzed in the EF43-*FGF4* mouse mammary tumor model, in which cells are implanted subcutaneously in immunocompetent BALB/c mice and reliably induce rapidly growing, highly vascularized tumors (Hajitou *et al.*, 2006b). Again a single systemic dose of RGD-4C AAVP-*HSVtk* followed by GCV markedly inhibited the growth of EF43-*FGF4* tumors. The RGD-4C AAVP-*HSVtk* vector was also evaluated in athymic rats (large rodents) bearing an orthotopic model of SKLMS human soft tissue sarcoma in the right hind limb and resulted in tumor growth suppression (Hajitou *et al.*, 2008). Histopathologic analysis of the tumors recovered 7 days after therapy with the targeted vector showed extensive destruction of the central area of the tumor with disrupted blood vessels and apoptosis and only a small viable outer rim with preserved vasculature and from which the tumors grew back (Hajitou *et al.*, 2006b, 2008).

Further to these experiments, researchers at the National Cancer Institute of the USA (NCI) have recently used the cancer targeting properties of RGD-4C AAVP to deliver the gene for the antitumor agent tumor necrosis factor α (TNFα) to the angiogenic vasculature of a human melanoma tumor model in nude mice (Tandle *et al.*, 2009). Systemic administration showed tumor-limited TNFα expression leading to induction of apoptosis in tumor blood vessels and significant inhibition of tumor growth without systemic toxicity to control organs (Tandle *et al.*, 2009).

Another recent study was also carried out under the direction of the NCI to assess the efficacy of targeted AAVP expressing the *TNFα* on dogs with spontaneous cancers (Paoloni *et al.*, 2009). They showed that intravenous single and multidoses of the vascular-targeted RGD-4C AAVP-*TNFα* were safe in dogs and led to selective localization of vector in the tumor blood vessels and specific expression of TNFα in tumors without any detectable TNFα in the control organs (Paoloni *et al.*, 2009). Miraculously, repeated vector administrations to a few dogs with very aggressive cancers, such as soft tissue sarcoma, resulted in complete eradication of the tumors and the dogs were cured (Paoloni *et al.*, 2009). Clinical trails in cancer patients are in plan and should happen in a relatively short time frame.

An important feature of the RGD-4C AAVP vector is that repeated administrations resulted in efficient antitumor therapy in immunocompetent mice and dogs despite the presence of high immune response against the bacteriophage (Hajitou *et al.*, 2006b; Paoloni *et al.*, 2009). Indeed, phage-based particles are known to be immunogenic, but this feature can be modulated through targeting itself (Hajitou *et al.*, 2006b).

E. Vascular-targeted systemic molecular-genetic imaging by RGD-4C AAVP

Previous reports have demonstrated some efficacy of cancer gene therapy in humans (Khuri et al., 2000; Nemunaitis et al., 2000). However, these successes are limited and the vast majority of clinical trials have failed to demonstrate any significant efficacy. Indeed, progress in the improvement of clinical cancer gene therapy has been hampered by difficulties in quantifying expression of the therapeutic transgene in vivo. In clinical trials of cancer gene therapy, this is normally achieved by analyzing biopsies from patients by molecular and histo-pathological methods. From these patient samples, the presence and expression of the transgene may be detected by several techniques: polymerase chain reaction (PCR), which is designed to amplify specifically the transgene; real-time PCR, which monitors the amplification of the transgene and allows the precise quantification of its expression level in a sample; histopathological methods, which identify the location of the transgene; or in some cases, mea-surement of the enzymatic activity of the transgene. However, the information that can be gathered from these types of approach is often limited in practice by tumor size and accessibility (Soghomonyan et al., 2005). Imaging is needed prior to gene therapy in order to assess the spatiotemporal expression and confirm that the therapeutic levels of transgene expression are achieved within the tumor tissue while the normal tissues are spared. Therefore, to ensure the rational development of gene therapy, a crucial issue is the utilization of technologies for the noninvasive monitoring of spatial and temporal transgene expression in vivo upon systemic administration of a gene delivery vector. Such imaging technologies would allow the generation of quantitative information about gene expression and the assessment of cancer gene therapy efficacy.

Noninvasive monitoring of gene delivery and expression is a very attractive approach since it can be repeated overtime in the same patient and on a whole body scale. Therefore, great effort is being invested in the field of molecular imaging, which can be defined as the visual representation, character-ization, and quantification of biological processes. The imaging of gene delivery and expression is increasingly becoming essential for cancer diagnosis, prediction of tumor response to current therapies, and monitoring response to therapies. In the past decade, progress has been made in the field of in vivo molecular-genetic imaging; however, meaningful results can only be achieved by an efficient targeted systemic gene delivery vector.

With the generation of the targeted AAVP, systemic molecular-genetic applications have become possible. AAVP efficacy was first assessed for preclini-cal bioluminescence imaging (BLI). BLI relies on light-generating enzymes (luciferases), which are used as reporters. The firefly luciferase enzyme, whose substrate is D-luciferin, emits light in the presence of oxygen and ATP, thus allowing the detection of gene expression *in vivo* (Weissleder, 2002). In addition, there is good, rapid distribution of D-luciferin *in vivo*. The firefly luciferase gene (*Luc*) is most commonly employed, other luciferases emit light at different wavelengths (e.g., *Renilla* luciferase, whose substrate is coelenterazine).

The targeted RGD-4C AAVP vector carrying the firefly luciferase reporter *Luc* gene was used for the imaging studies to target αv integrins over-expressed in tumor blood vessels (Hajitou *et al.*, 2006b) and to systemically and specifically deliver the firefly luciferase *Luc* reporter gene to tumors. Intravenous administration of the targeted vector to immunodeficient nude mice bearing subcutaneous human prostate xenograft DU145 led to a specific expression of luciferase in the tumor tissue without any detectable luciferase activity in the normal organs. AAVP was also tested in a model of athymic rnu/rnu nude rats bearing orthotopic human soft tissue sarcoma SKLMS (Fig. 3.1). Luciferase expression in tumors started at day 3 after intravenous administration of the targeted vector and increased gradually overtime to achieve maximum levels

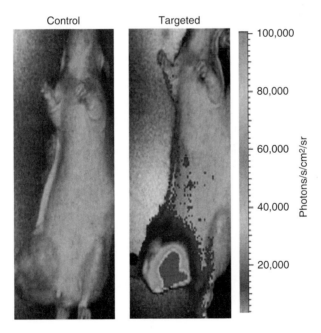

Figure 3.1. Bioluminescence imaging of *Luc* expression in tumor-bearing rats after intravenous administration of AAVP vectors. Nude rats bearing orthotopic human SKLMS soft tissue sarcoma xenografts in the right hind limb received a single intravenous dose (3×10^{12}TU), through the tail vein, of targeted RGD-4C AAVP or negative control nontargeted vectors carrying the firefly *Luc* reporter transgene. BLI of lucifearse is shown at day 34 after vector administration. A standard calibration scale is provided.

followed by a decrease of expression (Hajitou et al., 2008). All these results showed that targeted AAVP can be used in combination with BLI of luciferase for targeted systemic imaging of tumors in preclinical models.

F. Predicting therapy efficacy in cancer with vascular-targeted AAVP molecular imaging

Although BLI appears to be very effective in small animals, there is no evidence that BLI will be adaptable to larger animals. As for classic fluorescence imaging, light emitted from a deep source is greatly attenuated by surrounding tissues and cannot be detected. Therefore, BLI cannot currently be translated into clinical applications. However, the *HSVtk* gene can serve both as a "suicide" gene (when combined with GCV) and as a reporter transgene to test the clinical applicability of positron emission tomography (PET) imaging with *HSVtk*-specific radio-labeled nucleoside analogues such as ^{124}I-FAIU, ^{18}F-FHBG, or ^{18}F-FEAU. Detection in a PET scanner requires tracers that incorporate positron-emitting isotopes, such as ^{18}F or ^{124}I. Because PET allows quantification and high sensitivity, it appears to be the technique of choice to assess transgene expression upon *in vivo* delivery. It also constitutes a major advance in the assessment and follow-up of patients with cancer (Juweid and Cheson, 2006). In principle, PET directly measures the expression of a PET reporter protein and thereby gives an indirect evaluation of the expression of a therapeutic gene of interest. Small-animal PET devices have been developed. MicroPET (Hajitou et al., 2006b, 2008; Chatziioannou, 2002, 2005) is a noninvasive system that allows the acquisition of high-resolution 3D images from a small living animal (e.g., mice, rats). Scanning can be performed several times on the same living subject, allowing monitoring of the transgene expression and efficacy of a treatment. PET imaging of *HSVtk* expression provides the ability to define the location, magnitude, and duration of transgene expression. The radiolabeled nucleoside analogues which are specifically phosphorylated by HSVtk enzyme and get trapped inside the cell, have been used for the noninvasive localization of retroviral (Tjuvajev et al., 1996), herpes viral (Jacobs et al., 2001), and adenoviral (Gambhir et al., 2000) vector-mediated HSVtk expression in animal models.

The radiolabeled substrate [^{18}F]-FEAU was used in combination with microPET imaging to monitor the temporal dynamics and spatial heterogeneity of *HSVtk* gene expression in tumor-bearing animals after intravenous administration of the AAVP-*HSVtk* vector. This component is regarded as an excellent substrate for the HSVtk enzyme because of the very low background activity in all normal organs and tissues (Kang et al., 2005). Mice bearing DU145-derived tumor xenografts were administered with a single dose of AAVP vector intravenously, then given [^{18}F]-FEAU intravenously at different time points in order to carry out repetitive PET imaging. A gradual increase of [^{18}F]-FEAU, thus of the level of *HSVtk* transgene expression in tumors was observed during the first 5 days after administration of the targeted RGD-4C AAVP-*HSVtk*, followed by gradual stabilization of *HSVtk* expression by day 10 (Hajitou *et al.*, 2006b). Only background levels of [^{18}F]-FEAU were observed in mice that received control nontargeted vectors (Hajitou *et al.*, 2006b). GCV treatment was initiated in mice when a plateau of *HSVtk* expression was achieved in tumors. Mice were treated daily with GCV, and tumor viability was monitored with an intravenous injection of the metabolic tracer ^{18}F-labeled fluorodeoxyglucose ([^{18}F]-FDG) and serial PET imaging (Hajitou *et al.*, 2006b). FDG is a glucose analogue that is phosphorylated by the cells and trapped as FDG-6-phosphate inside the cells. The images showed that the tumors were viable and actively accumulated [^{18}F]-FDG before GCV treatment, while significantly regressed and metabolically suppressed (decrease in accumulation of [^{18}F]-FDG) after GCV injection to mice that received the targeted RGD-4C AAVP-*HSVtk* vector (Hajitou *et al.*, 2006b).

These results showed that [^{18}F]-FEAU can be used with PET to detect cells expressing the *HSVtk* after systemic delivery of targeted AAVP, indicating that AAVP-mediated PET scan of *HSVtk* expression enables molecular prediction of drug response in preclinical models of cancer.

Similar results were obtained by using the orthotopic model of human soft tissue sarcoma xenografts in nude rats. The nucleoside analogue [^{18}F]-FEAU and microPET imaging was used to repetitively monitor the location, magnitude, and temporal dynamics of *HSVtk* expression after intravenous administration of targeted AAVP-*HSVtk* (Hajitou *et al.*, 2008). The targeted AAVP enabled monitoring of tumor drug response in this preclinical experimental model of human soft tissue sarcoma xenografts in nude rats. Specifically, intravenous administration of the targeted AAVP-*HSVtk* allowed noninvasive serial imaging of the reporter/therapeutic *HSVtk* transgene and prediction of tumor response to GCV with PET-based imaging of RGD-4C AAVP-delivered *HSVtk* incorporating specific radiotracers.

Such vascular-targeted molecular-genetic imaging provides a potential advantage because of its ability to monitor the efficiency and specificity of *HSVtk* delivery before and to predict tumor response after GCV treatment. Our proof of concept results suggest that targeted AAVP can be integrated with PET imaging to predict tumor response to therapy.

III. CONCLUSIONS AND PERSPECTIVES

The vascular endothelium is an important target for therapeutic diagnostic interventions in several human diseases including cancer. Blood stream is a natural route for vascular delivery of gene therapy vectors. However, efficacy has been hindered by the lack of specificity of vectors and their uptake by the liver and reticuloendothelial system. Bacteriophages have distinct advantages over existing gene therapy vectors because they are simple, economical to produce at high titer, have no intrinsic tropism for mammalian cells, and are relatively simple to genetically modify, evolve, and target to the tissues of interest.

The AAVP-based vector system has potential biological and clinical applications including targeted gene transfer and molecular-genetic imaging. Vascular targeting of cancer after systemic delivery of AAVP vectors has already been established. To date, the reports continue to show that AAVP can be used to target gene therapy and molecular imaging to distant tumors after systemic administration, a route that is applicable for localized and metastatic disease. Tissue-targeted AAVP particles can also be developed for targeted systemic delivery of therapeutic and imaging genes to normal tissues. The *in vivo* phage display technology is thus a powerful tool to be used to isolate and identify phage clones able to target, diffuse, and achieve internalization in the normal tissues after systemic administration. The selected phage clone displaying a specific peptide can be directly used to generate a tissue-targeted AAVP vector for systemic gene delivery. Moreover, targeted molecular imaging is a promising approach to study the persistence of transgene expression in tumors after systemic administration of targeted AAVP, and also provides the ability to visualize and quantify not only the magnitude and spatial distribution of transgene expression, but also the temporal dynamics (with repetitive molecular imaging). The use of targeted AAVP carrying transgenes that serve dual reporter/therapeutic genes such as *HSVtk* could provide a clinic-ready AAVP-based candidate approach for translation in human cancer and lead to a personalized cancer gene therapy.

References

Arap, W., Pasqualini, R., and Ruoslahti, E. (1998). Cancer treatment by targeted drug delivery to tumor vasculature in a mouse model. *Science* **279,** 377–380.

Arap, W., Kolonin, M. G., Trepel, M., *et al.* (2002). Steps toward mapping the human vasculature by phage display. *Nat. Med.* **8,** 121–127.

Barrow, P. A., and Soothill, J. S. (1997). Bacteriophage therapy and prophylaxis: Rediscovery and renewed assessment of potential. *Trends Microbiol.* **5,** 268–271.

Bertoni, C., Jarrahian, S., Wheeler, T. M., *et al.* (2006). Enhancement of plasmid mediated gene therapy for muscular dystrophy by directed plasmid integration. *Proc. Natl. Acad. Sci. USA* **103,** 419–424.

Bouard, D., Alazard-Dany, N., and Cosset, F. L. (2009). Viral vectors: From virology to transgene expression. *Br. J. Pharmcol.* **157,** 153–165.

Brooks, P. C., Montgomery, A. M., Rosenfeld, M., *et al.* (1994). Integrin $\alpha_v\beta3$ antagonists promote tumor regression by inducing apoptosis of angiogenic blood vessels. *Cell* **79,** 1157–1164.

Büning, H., Perabo, L., Coutelle, O., *et al.* (2008). Recent developments in adeno-associated virus vector technology. *J. Gene Med.* **10,** 717–733.

Chatziioannou, A. F. (2002). PET scanners dedicated to molecular imaging of small animal models. *Mol. Imaging Biol.* **4,** 47–63.

Chatziioannou, A. F. (2005). Instrumentation for molecular imaging in preclinical research: Micro-PET and Micro-SPECT. *Proc. Am. Thorac. Soc.* 2(533–536), 510–511.

Dickerson, E. B., Akhtar, N., Steinberg, H., *et al.* (2004). Enhancement of the antiangiogenic activity of interleukin-12 by peptide targeted delivery of the cytokine to $\alpha_v\beta3$ integrin. *Mol. Cancer Res.* **2,** 663–673.

Douglas, J. T. (2007). Adenoviral vectors for gene therapy. *Mol. Biotechnol.* **36,** 71–80.

Duan, D., Sharma, P., Yang, J., *et al.* (1998). Circular intermediates of recombinant adeno-associated virus have defined structural characteristics responsible for long-term episomal persistence in muscle tissue. *J. Virol.* **72,** 8568–8577.

Ellerby, H. M., Arap, W., Ellerby, L. M., *et al.* (1999). Anti-cancer activity of targeted pro-apoptotic peptides. *Nat. Med.* **5,** 1032–1038.

Folkman, J. (1997). Principles and Practice of Oncology. *In* "Cancer" (V. T. DeVita, S. Hellman, and S. A. Rosenberg, eds.), 5th edn., pp. 3075–3085. Lippincott, Philadelphia.

Gambhir, S. S., Bauer, E., Black, M. E., *et al.* (2000). A mutant herpes simplex virus type 1 thymidine kinase reporter gene shows improved sensitivity for imaging reporter gene expression with positron emission tomography. *Proc. Natl. Acad. Sci. USA* **97,** 2785–2790.

Gosh, D., and Barry, M. A. (2005). Selection of muscle-binding peptides from context-specific peptide-presenting phage libraries for adenoviral vector targeting. *J. Virol.* **79,** 13667–13672.

Green, N. K., and Seymour, L. W. (2002). Adenoviral vectors: Systemic delivery and tumor targeting. *Cancer Gene Ther.* **9,** 1036–1042.

Gregorevic, P., Blankinship, M. J., Allen, J. M., *et al.* (2004). Systemic delivery of genes to striated muscles using adeno-associated viral vectors. *Nat. Med.* **10,** 828–834.

Hajitou, A., Pasqualini, R., and Arap, W. (2006a). Vascular targeting: Recent advances, and therapeutic perspectives. *Trends Cardiovasc. Med.* **16,** 80–88.

Hajitou, A., Trepel, M., Lilley, C. E., *et al.* (2006b). A hybrid vector for ligand-directed tumor targeting and molecular imaging. *Cell* **125,** 385–398.

Hajitou, A., Rangel, R., Trepel, M., *et al.* (2007). Design and construction of targeted AAVP vectors for mammalian cell transduction. *Nat. Protoc.* **2,** 523–531.

Hajitou, A., Levy, D. C., Hannay, J. A. F., *et al.* (2008). A preclinical model for predicting drug response in soft-tissue sarcoma with targeted AAVP molecular imaging. *Proc. Natl. Acad. Sci. USA* **105,** 4471–4476.

Hendrix, R. W. (1999). Evolution: The long evolutionary reach of viruses. *Curr. Biol.* **9,** R914–R917.

Hida, K., and Klagsburn, M. (2005). A new perspective on tumor endothelial cells: Unexpected chromosome and centrosome abnormalities. *Cancer Res.* **65,** 2507–2510.

Hood, J. D., Bednarski, M., Frausto, R., *et al.* (2002). Tumor regression by targeted gene delivery to the neovasculature. *Science* **296,** 2404–2407.

Jacobs, A., Tjuvajev, J. G., Dubrovin, M., *et al.* (2001). Positron emission tomography-based imaging of transgene expression mediated by replication-conditional, oncolytic herpes simplex virus type 1 mutant vectors in vivo. *Cancer Res.* **61,** 2983–2995.

Juweid, M. E., and Cheson, B. D. (2006). Positronemission tomography and assessment of cancer therapy. *N. Engl. J. Med.* **354,** 496–507.

Kanerva, A., Zinn, K. R., Peng, K. W., *et al.* (2005). Noninvasive dual modality in vivo monitoring of the persistence and potency of a tumor targeted conditionally replicating adenovirus. *Gene Ther.* **12,** 87–94.

Kang, K. W., Min, J. J., Chen, X., and Gambhir, S. S. (2005). Comparison of [^{14}C]FMAU, [^{3}H] FEAU, [^{14}C]FIAU, and [^{3}H]PCV for monitoring reporter gene expression of wild type and mutant *Herpes simplex* virus type 1 thymidine kinase in cell culture. *Mol. Imaging Biol.* **7,** 296–303.

Khuri, F. R., Nemunaitis, J., Ganly, I., et al. (2000). A controlled trial of intratumoral ONYX-015, a selectively-replicating adenovirus, in combination with cisplatin and 5-fluorouracil in patients with recurrent head and neck cancer. Nat. Med. **6,** 879–885.

Larocca, D., Witte, A., Johnson, W., et al. (1998). Targeting bacteriophage to mammalian cell surface receptors for gene delivery. Hum. Gene Ther. **9,** 2393–2399.

Marchiò, S., Lahdenranta, J., Schlingemann, R. O., et al. (2004). Aminopeptidase A is a functional target in angiogenic blood vessels. Cancer Cell **5,** 151–162.

Meykadeh, N., Mirmohammadsadegh, A., Wang, Z., et al. (2005). Topical application of plasmid DNA to mouse and human skin. J. Mol. Med. **83,** 897–903.

Nemunaitis, J., Swisher, S. G., Timmons, T., et al. (2000). Adenovirus-mediated p53 gene transfer in sequence with cisplatin to tumors of patients with non-small-cell lung cancer. J. Clin. Oncol. **18,** 609–622.

Oosterhoff, D., Pinedo, H. M., Witlox, M. A., et al. (2005). Gene-directed enzyme prodrug therapy with carboxylesterase enhances the anticancer efficacy of the conditionally replicating adenovirus AdDelta24. Gene Ther. **12,** 1011–1018.

Paoloni, M., Tandle, A., Mazcko, C., et al. (2009). Launching a Novel Preclinical Infrasturcture: Comparative Oncology Trials Consortium Directed Therapeutic Targeting of TNF-α to Cancer Vasculature. PLoS One **4,** e4972.

Pasqualini, R., and Ruoslahti, E. (1996). Organ targeting in vivo using phage display peptide libraries. Nature **380,** 364–366.

Pasqualini, R., Koivunen, E., Kain, R., et al. (2000). Aminopeptidase N is a receptor for tumor-homing peptides and a target for inhibiting angiogenesis. Cancer Res. **60,** 722–727.

Peek, R., and Reddy, K. R. (2006). FDA approves use of bacteriophages to be added to meat and poultry products. Gastroenterology **131,** 1370–1372.

Sadeghi, H., and Hitt, M. M. (2005). Transcriptionally targeted adenovirus vectors. Curr. Gene Ther. **5,** 411–427.

Samulski, R. J., Chang, L. S., and Shenk, T. (1987). A recombinant plasmid from which an infectious adeno-associated virus genome can be excised in vitro and its use to study viral replication. J. Virol. **61,** 3096–3101.

Soghomonyan, S., Doubrovin, M., Pike, J., et al. (2005). Positron emission tomography (PET) imaging of tumor-localized Salmonella expressing HSV1-TK. Cancer Gene Ther. **12,** 101–108.

Tandle, A., Hanna, E., Lorang, D., et al. (2009). Tumor vasculature targeted delivery of tumor necrosis factor-alpha. Cancer **115,** 128–139.

Tjuvajev, J. G., Finn, R., Watanabe, K., et al. (1996). Noninvasive imaging of herpes virus thymidine kinase gene transfer and expression: A potential method for monitoring clinical gene therapy. Cancer Res. **56,** 4087–4095.

Trepel, M., Arap, W., and Pasqualini, R. (2002). In vivo phage display and vascular heterogeneity: Implications for targeted medicine. Curr. Opin. Chem. Biol. **6,** 399–404.

Trepel, M., Stoneham, C. A., Eleftherohorinou, H., et al. (2009). A heterotypic bystander effect for tumor cell killing after AAVP-mediated vascular targeted suicide gene transfer. Mol. Cancer Ther. **8,** 2383–2391.

Waehler, R., Russell, S. J., and Curiel, D. T. (2007). Engineering targeted viral vectors for gene therapy. Nat. Rev. Genet. **8,** 573–587.

Wang, Z., Zhu, T., Qiao, C., et al. (2005). Adeno-associated virus serotype 8 efficiently delivers genes to muscle and heart. Nat. Biotechnol. **23,** 321–328.

Wang, Z., Kuhr, C. S., Allen, J. M., et al. (2007). Sustained AAV-mediated dystrophin expression in a canine model of Duchenne muscular dystrophy with a brief course of immunosuppression. Mol. Ther. **15,** 1160–1166.

Weissleder, R. (2002). Scaling down imaging: Molecular mapping of cancer in mice. Nat. Rev. Cancer **2,** 11–18.

Cationic and Tissue-Specific Protein Transduction Domains: Identification, Characterization, and Therapeutic Application

4

Maliha Zahid, Xiaoli Lu, Zhibao Mi, and Paul D. Robbins

Department of Microbiology and Molecular Genetics, 427 Bridgeside Point II, 450 Technology Drive, University of Pittsburgh School of Medicine, Pittsburgh, Pennsylvania, USA

I. Introduction
II. Cationic Protein Transduction Domains
III. Biopanning using Peptide Phage Display Libraries
IV. Peptide Phage Display Biopanning in Cell Culture for Transduction Peptides
V. Methods for Analysis of Transduction Efficiency
VI. Peptide Phage Display Biopanning *in vivo* for Transduction Peptides
VII. PTDs as Therapeutics
VIII. Conclusions
References

ABSTRACT

Protein transduction domains (PTDs) are small peptides able to transverse plasma membranes, able to carry proteins, nucleic acid, and viral particles into cells. PTDs can be broadly classified into three types; cationic, hydrophobic, and cell-type specific. The cationic PTDs, comprised of arginines, lysines, and ornithines, and hydrophobic PTDs can efficiently transduce a variety of cell types in culture and *in vivo*. The tissue-specific transduction domains, identified by screening of peptide display phage libraries for peptides able to confer internalization, have more restricted transduction properties. Here we provide a

Advances in Genetics, Vol. 69
Copyright 2010, Elsevier Inc. All rights reserved.

0065-2660/10 $35.00
DOI: 10.1016/S0065-2660(10)69007-4

review of PTDs, focusing on methods for identifying and characterizing both cationic and tissue-specific transduction peptides. In particular, we describe the use of screening peptide phage display libraries to identify tissue-specific transduction peptides. © 2010, Elsevier Inc.

I. INTRODUCTION

Protein transduction domains (PTDs) or cell penetrating peptides (CPPs) are small peptides able to transverse plasma membranes and carry full-length proteins, oligonucleotides, iron nanoparticles, and liposomes as "cargoes." These PTDs can be broadly classified under three classes; (1) cationic or positively charged PTDs, (2) hydrophobic or protein leader sequence-derived domains, and (3) peptides identified by phage display that are able to transduce cells in a cell-type-specific manner. Interest in these types of peptides, in particular, the cationic PTDs, began with the simultaneous reports by Frankel and Pabo (1988) and Green and Loewenstein (1988), that HIV-1 TAT protein was able to cross plasma membranes. Similarly, the Antennapedia homeodomain of *Drosophila* was shown to cross cell membranes efficiently (Joliot *et al.*, 1991). Further investigation identified small domains within these full-length proteins as being responsible for their transduction abilities. These small domains, termed PTDs, are able to carry large complexes as cargoes across cell membranes (Allinquant *et al.*, 1995; Fawell *et al.*, 1994; Schwarze *et al.*, 1999). The interest in these unique peptides has led to the identification of multiple naturally occurring (Derossi *et al.*, 1994; Morris *et al.*, 1997; Pooga *et al.*, 1998) as well as engineered PTDs (Futaki *et al.*, 2001; Mai *et al.*, 2002).

II. CATIONIC PROTEIN TRANSDUCTION DOMAINS

Both Antp and TAT PTDs are enriched for cationic amino acids (arginines and lysines), which play a critical role in mediating transduction. We and others have shown that synthetic peptides comprised of 6–10 arginines, lysines, or ornithines also work as potent transduction domains. These positively charged particles interact with negatively charged membrane components, such as glycosaminoglycans, to initiate the first steps in transduction. Indeed, we have demonstrated that cells deficient in glycosaminoglycan synthesis are transduced with a lower efficiency by cationic PTD, but the transduction efficiency can be increased by treating the glycosaminoglycan-deficient cells with dextran sulfate. Several mechanisms have been suggested to explain transduction of PTDs following binding to the cell including macropinocytosis associated with lipid rafts. However, it is likely that different cationic peptides transduce cells via different mechanisms or that the same peptide transduces through multiple mechanisms. The fact that transduction can occur via both energy independent and dependent manners suggest that there may be at least two separate mechanisms for transduction.

Multiple studies have demonstrated the ability of cationic PTDs to deliver peptides of therapeutic potential (Choi et al., 2003; Dave et al., 2007), full-length proteins (Ribeiro et al., 2003; Schwarze et al., 1999), oligonucleotide (Eguchi et al., 2001), 40 nm iron nanoparticles (Josephson et al., 1999; Lewin et al., 2000), drugs (Rothbard et al., 2000), and liposomes (Torchilin et al., 2001). In addition, there is extensive interest in developing PTDs as vehicles for delivery of small interfering RNA (siRNA) in culture (Chiu et al., 2004; Muratovska and Eccles, 2004) and in vivo (Kumar et al., 2007; Moschos et al., 2007), with novel strategies aimed at PTD and siRNA complex escaping the endosomal compartment (Endoh et al., 2008; Lundberg et al., 2007) or increasing efficacy by avoiding the nuclear compartmentalization (Lundberg et al., 2007). These numerous applications for PTDs highlight the potential therapeutic applications of PTDs for delivery of myriad agents for treatment of different diseases. Although PTDs have shown the ability to deliver biologically active cargos in vivo and even cross the blood–brain barrier (Schwarze et al., 1999), the lack of cell or tissue specificity observed with the cationic or hydrophobic PTDs could increase their toxicity and thus limits their use in vivo. Thus, there is a need, at least for certain therapeutic applications, to identify transduction peptides that are tissue specific in order to limit toxicity and enhance therapeutic efficacy. To identify tissue-specific transduction peptides, biopanning or phage display has emerged as an effective approach for screening for peptides. Indeed, we have used this approach to identify tissue-specific transduction peptides that are functional in vivo.

III. BIOPANNING USING PEPTIDE PHAGE DISPLAY LIBRARIES

The principal underlying display technologies consists of the ability to physically link phenotypes of polypeptides displayed on a certain platform, for example bacteriophage, to their corresponding genotype. Typically, the phenotype registered in display experiments is through physical binding and serial selection. The ability to modify filamentous bacteriophage to express polypeptides on the surface as part of the bacteriophage's coat protein was first reported by Smith (1985). This technique was originally developed to map epitope-binding sites of antibodies by mapping against immobilized targets. This seminal work led to phage display being utilized as a powerful method to identify specific peptide binding with a diverse range of biological as well as technical and medical applications (Scott and Smith, 1990).

Bacteriophages (phages) are prokaryotic bacterial viruses, which are the most abundant biological entities in the environment with estimates ranging from 10^{30} to 10^{32} and play a key role in many biological systems. Bacteriophages as a group are extremely diversified. The number of known phage has been expanding for decades at a rate of ~100 per year. Phages are broadly classified as tailed, polyhedral, filamentous, or pleomorphic. According to the latest electron microscopic analysis of over 500 phages (Ackermann, 2007), the vast majority (~96%) are tailed, with polyhedral, filamentous, and pleomorphic phages comprising 3.7%. Of these, filamentous phage has been studied widely for the purpose of phage display. Filamentous phage is nonlytic and leaves the host by budding through the host plasma membrane without killing the host, though markedly slowing the rate of replication. The filamentous phage best studied biochemically and genetically is the M13 and its close relative fd bacteriophage. Both are F-specific phages that infect *Escherichia coli* bacteria.

M13 phage is a long, thread shaped particle, 6.5 nm in diameter, and 900 nm in length. Its genome is a single-stranded, circular DNA containing 6407 nucleotides protected by a long cylindrical protein coat. The most abundant coat protein is the major coat protein pVIII, of which ~2700 molecules are present on the coat. There are only five copies of the minor coat protein pIII at one end of the phage particle, which along with five molecules of another minor coat protein, pVI, are involved in bacterial cell binding. At the other end of the phage are minor coat proteins pVII and PIX, which are needed for initiation and maintenance of phage assembly inside the host. During the assembly process the resulting fusion coat proteins are transported to the bacterial periplasm or inner cell membrane and incorporated into a new phage particle along with the single-stranded DNA genome of the phage carrying the genotypic information for the displayed fusion protein.

Libraries have been constructed utilizing virtually all of the coat proteins of filamentous phage being modified for the purpose of phage display. However, the most common filamentous phage libraries consist of phage with modified minor capsid protein pIII, expressing 10^9 peptides or more of different permutations. Filamentous phage infects pilus-positive bacteria that are not lysed by the infecting phage and secretes multiple, identical copies of it displaying a particular insert. Phage display constitutes exposing the target of interest to a large, randomized library of phage expressing peptides on a surface coat protein. After binding, the nonrelevant phage is washed away, the bound phage eluted, expanded in pili-positive host bacteria, and reexposed to the target of interest. Four to six cycles of biopanning lead to adequate enrichment with the relevant clone phage, which can subsequently be sequenced, and the peptide motif carried as a fusion coat protein identified.

Initial studies utilizing phage display libraries were limited to *in vitro* studies, phage display being carried out against immobilized antigen targets or cell-specific ligands. Arap *et al.* (1998) demonstrated the feasibility of *in vivo* phage display leading to identification of peptides homing to tumor vasculature as well as the ability to use these peptides to target anticancer therapy. Further *in vivo* work has clarified that not only tumor vasculature, but also normal organs, like adipose tissue (Kolonin *et al.*, 2004) and the heart (Zhang *et al.*, 2005), among other organs (Arap *et al.*, 2002; Pasqualini *et al.*, 2001), bear ligands that can be targeted by *in vivo* phage display. However, these studies focused on identifying peptides able to bind to the surface of the cell. Although some of the peptide targets are receptors that are internalized, the screening approach was not developed to identify transduction peptide specifically. We have modified the protocols for phage display biopanning to screen for peptides able to bind and be internalized into the specific types of cells. Using this modified approach we have identified peptides able to transduce synoviocytes, fibroblasts which line the joint space (Mi *et al.*, 2003), murine tumor cells, airway epithelia cells, and cardiomyoctes in culture and *in vivo*.

IV. PEPTIDE PHAGE DISPLAY BIOPANNING IN CELL CULTURE FOR TRANSDUCTION PEPTIDES

We have developed a biopanning methodology to identify peptides able to facilitate transduction of specific cell types using peptide phages display. We have used this approach to identify synovial fibroblast, airway epithelial, and tumor-specific and cardiac-tissue-specific transduction peptides. An example of the methodology we have used is described below for the identification of an epithelial cell-specific transduction peptide. For this, we screened an M13 phage 12-amino acid peptide display library for peptides able to transduce HEK293 cells in culture. The 293 cells were incubated with 4×10^{10} phage for 3 h at 37 °C or overnight at 4 °C. After the incubation period, the cells were washed extensively (20 times) with Tris–NaCl buffer, recovered by trypsinization and then lysed by three consecutive rounds of freeze–thaw. Cell debris was pelleted out and the phage eluted with glycine, tittered, and amplified for a second round of biopanning. A total of four cycles were performed. Following the last cycle of biopanning, 20 plaques were picked and the phage amplified. To determine if the identified phage were indeed being internalized into 293 cells, each isolated phages was labeled with a fluorescent label, Cy3 (Sigma, St. Louis, MO) as described (Kelly *et al.*, 2006). The phage–Cy3 complex was then added to the media of 293 cells for 3 h, after which time the cells were washed extensively, fixed with 4% paraformaldehyde, nuclei counterstained with Sytox green, and the cells imaged with confocal fluorescent microscopy. There was differential uptake of the labeled phage by 293 cells with phage #12 showing the most robust transduction with lesser extent of transduction occurring with phage #10 and #15 (Fig. 4.1).

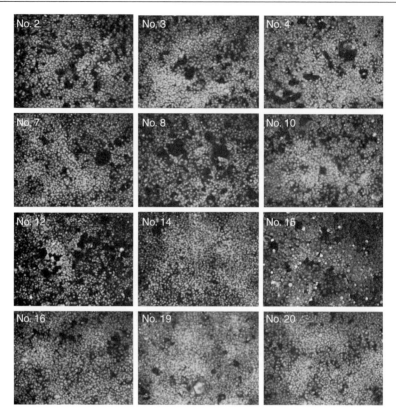

Figure 4.1. Cy3-labeled phage internalization into 293 cells. (See Color Insert.)

From the above set of experiments it was clear that phage #12 showed the most robust internalization by 293 cells. Thus this phage was amplified, DNA extracted, and the encoded peptide sequenced using the primers provided with the phage display library. The identified peptide sequence, termed 293P (SNNNVRPIHIWP), was synthesized coupled to 6CF. Multiple different cell lines, including the epithelial cell lines HeLa, H1299, and HEK293 cells, were incubated with 200 μM concentration of the 293P, a cationic transduction peptide PTD-5, or a control peptide (Con-P). 293P was able to transduce HEK293 cells, as well as to a lesser extent HeLa and H1299 cells, with no transduction of other cell types observed. In contrast, the cationic transduction domain PTD-5 was taken up by all cell types, to varying extents, in a non-cell-type-specific manner. There was no appreciable cellular uptake of Con-P. Given that 293T, HeLa, and H1299 are all epithelial cell lines, the transduction results with 293P suggest that it can transduce epithelial cells, at least in culture.

Analysis of transduction of 293 cells at 37 and 4 °C revealed that the mechanism of transduction is different between the tissue-specific 293P and the cationic PTD-5 (Figs. 4.2 and 4.3). Transduction by 293P was significantly reduced at 4 °C, whereas there was only a minimal effect on transduction by PTD-5.

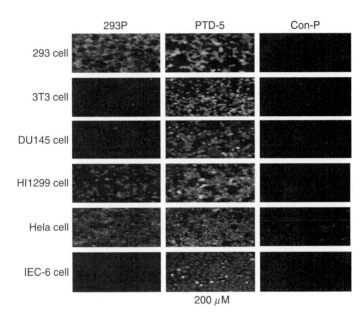

Figure 4.2. Cell specificity of 293P identified from biopanning of a M13 peptide phage display library. (See Color Insert.)

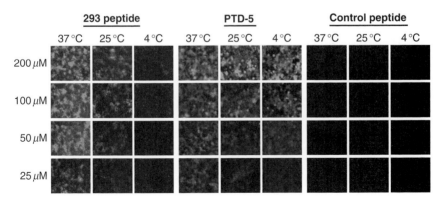

Figure 4.3. Reduced transduction by 293P at 4 °C. (See Color Insert.)

Presumably 293P is interacting with a specific membrane component present on epithelial cells such as HEK293, HeLa, and H1299 which is subsequently internalized. In contrast, PTD-5 interacts with general components of the membranes such as glycosaminoglycans, in particular heparin sulfate, followed by internalization through one or more different pathways.

The experiments described above illustrate, as other studies before them (Mai *et al.*, 2001; May *et al.*, 2000; Molenaar *et al.*, 2002; Strickland and Ghosh, 2006; Tilstra *et al.*, 2007), the feasibility of phage display to identify peptides able to transduce specific cell types without prior knowledge of the cell membrane component being targeted. Not only can peptides targeting specific cell types be identified using this technique, but also phage bearing these ligands can be internalized by the cell being targeted demonstrating the ability of these peptides to act as "cargo-carriers." Further study of these peptides and their binding partners could lead to important insights into cellular physiology and potentially lead to better targeting agents.

V. METHODS FOR ANALYSIS OF TRANSDUCTION EFFICIENCY

An important issue regarding the characterization of transduction peptides is clearly documenting functional transduction. The majority of the studies to examine transduction were performed using the peptide coupled or conjugated to a fluorescent marker like 6-carboxyfluorescein, streptavidin-alexa488, eGFP, or β-galactosidase. The analysis of fluorescence or enzymatic activity in the cells or in cell extracts can provide strong evidence for internalization as well as provide evidence for intracellular localization. However, to demonstrate functional delivery of agents into the cell, we have used two different approaches. One approach involves the use of a peptide inhibitor of the transcription factor NF-κB, a ubiquitously expressed transcription factor that is evolutionarily conserved (Hayden and Ghosh, 2008). The majority of NF-κB is localized in the cytoplasm, due to its binding to IκB inhibitor proteins. NF-κB activation occurs through the IκB kinase (IKK), a large 700–900 kDa complex containing two catalytic subunits, IKKα and IKKβ, and a regulatory subunit termed IKKγ or NEMO (NF-κB essential modulator). In response to a variety of stimuli, which include proinflammatory cytokines, bacterial products, viruses, and growth factors, IKK becomes activated and phosphorylates IκB, leading to its ubiquitination and subsequent degradation by the 26S proteasome. IκB degradation allows NF-κB to translocate to the nucleus where it binds to its cognate DNA site to induce gene expression. A small domain (11 amino acids) has been identified within IKKβ, termed the NEMO binding domain (NBD), which confers binding to IKKγ (May *et al.*, 2000). Delivery of the NBD using a PTD results in inhibition of the interaction of IKKβ with its regulatory subunit IKKγ. In particular, when this short peptide NBD (TALDWSWLQTE) is linked to a PTD, it leads to a dose dependent inhibition of NF-κB signaling in tissue culture and in animal models. Thus we have used PTD-mediated delivery of NBD and inhibition of an NF-κB reporter to assess and compare efficacy of functional transduction.

A second approach has been through the use of PTD-mediated delivery of an antimicrobial peptide to cells. Many antimicrobial peptides work by disrupting the bacterial membrane, but have no effect on eukaryotic membranes. However, these anti-microbial peptides can disrupt the integrity of mitochondrial membranes, resulting in release of cytochrome C and induction of apoptosis. PTD-mediated delivery of a well-characterized antimicrobial peptide, KLAKLAK, results in rapid and efficient disruption of the mitochondrial membrane and induction of apoptosis (Mai et al., 2001). However, without fusion to a PTD, the KLAK peptide has no effect when added to cells in culture. Thus we have used induction of apoptosis by the KLAKLAK peptide, fused to PTDs, as an indicator of functional transduction of cells. For example, fusion of the synovial-specific transduction domain to the KLAK peptide resulted in induction of apoptosis of synovial fibroblasts specifically in culture and in vivo (Mi et al., 2003). Similarly, fusion of a cationic PTD to KLAKLAK resulted in a peptide that could induce rapid induction of apoptosis in tumor cells following direct intratumoral injection (Mai et al., 2001).

VI. PEPTIDE PHAGE DISPLAY BIOPANNING *IN VIVO* FOR TRANSDUCTION PEPTIDES

In addition to cell culture approaches for peptide phage display biopanning to identify transduction peptides, the screening can also be performed in vivo. In vivo phage display is complicated by the complexity of the binding target, a large pool of nonspecific, contaminating phage and robust uptake of the phage by liver and other organs of the reticuloendothelial system (Molenaar et al., 2002). Taking cues from the biodistribution work done by Molenaar and colleagues (Molenaar et al., 2002), where the half-life of M13 phage was shown to be 4.5 h, the problem of the large pool of contaminating, nonspecific phage in the intravenously injected phage can be minimized by allowing for longer circulation times. This would allow bound phage to be retained with clearance of nonspecific phage from the circulation, largely by the liver. Alternatively, in vivo phage display can be utilized after a first prescreening cycle of phage display carried out on relevant cell line to enrich the pool for phage targeting the organ of interest. For the identification of internalizing peptides, we developed a protocol where the mice were pretreated with chloroquine (20 mg/kg) 24 h prior to and on the day of the intravenous phage injection. Chloroquine treatment was used to minimize intralysosomal destruction of internalized phage and theoretically increase the chances of recovering internalized phage. The phage was allowed to circulate for 24 h when the mice were euthanized and heart and kidney tissues obtained. The collected tissues were digested with collagenase and phage recovered by a single freeze/thaw cycle. Recovered phage was then tittered, normalized by tissue weight and subsequently amplified for subsequent rounds of biopanning. Utilizing such a combinatorial approach of screening in cell culture followed by *in vivo* screening, we have identified a peptide able to target H9C2 cells, a rat cardiomyoblast cell line *in vitro*, and mouse heart tissue *in vivo*, in a tissue-specific manner.

VII. PTDS AS THERAPEUTICS

Identifying peptides able to act as tissue-specific or general PTDs are likely to enhance the diagnostic as well as the therapeutic potential of this ever-expanding repertoire of PTDs. For example, we and others have used the NBD peptide described above not only for assessment of transduction efficiency, but also as a therapeutic agent. Dysregulation of the IKK/NF-κB pathway is frequently seen in several pathophysiological states including cancer, rheumatoid arthritis, sepsis, asthma, muscular dystrophy, multiple sclerosis, atherosclerosis, heart disease, inflammatory bowel disease, bone resorption, and diabetes, both type I and II. Initially it was shown that NBD was able to inhibit carrageenan induced paw swelling (May et al., 2000; Tilstra et al., 2007), similar to our recent observation that local injection of NBD is able to inhibit footpad swelling in a murine model of delayed type hypersensitivity (Tilstra et al., 2007). The NBD peptide was also shown to inhibit progression of collagen-induced arthritis, preventing paw and joint swelling with less bone and cartilage degradation when administered prior to and even after disease onset (Strickland and Ghosh, 2006). The NBD peptide also has been used to treat a murine model of multiple sclerosis (EAE), improving clinical symptoms and shifting the immune response from a Th1 to a Th2 profile (Dasgupta et al., 2004). Interestingly, systemic treatment of a murine model of muscular dystrophy, the *mdx* mouse, with Antp-NBD resulted in significant improvement in muscle pathology with less muscle inflammation and degeneration, and enhanced muscle regeneration (Acharyya et al., 2007). In addition, we observed significant therapeutic effects of the 8K-NBD peptide in a murine model of inflammatory bowel disease (Dave et al., 2007). Finally, it was demonstrated recently that the Antp-NBD peptide was highly therapeutic when given systemically in a mouse model of Parkinson's disease (Ghosh et al., 2007), suggesting that the peptide can cross the blood–brain barrier and block neurodegeneration. Taken together, these results suggest that the NBD peptide, when delivered using a PTD, not only blocks inflammatory and autoimmune diseases, but also has significant effects on tissue regeneration.

VIII. CONCLUSIONS

PTDs are small peptides able to deliver therapeutic agents efficiently across cell membranes. Although cationic transduction domains were first identified in naturally occurring proteins such as HIV TAT and *Drosophila* Antennapedia proteins, synthetic cationic peptides have been developed that are even more potent for transduction of cells. Tissue-specific transduction peptides can be identified by screening peptide phage display libraries, both in cell culture and *in vivo*, for internalized phage. These tissue-specific transduction peptides presumably transduce cells through a different mechanism than cationic PTDs, interacting with tissue-specific membrane components. Both the tissue-specific and cationic peptides have been used therapeutically in animal models. In particular, the PTD-mediated delivery of a peptide inhibitor of IKK/NF-κB signaling, NBD, has shown significant therapeutic effects in murine models of degenerative, autoimmune, and inflammatory diseases. Clearly PTD-mediated delivery of peptides, full-length proteins, nucleic acid, and drugs will be useful for the treatment of many different types of diseases.

References

Acharyya, S., *et al.* (2007). Interplay of IKK/NF-kappaB signaling in macrophages and myofibers promotes muscle degeneration in Duchenne muscular dystrophy. *J. Clin. Invest.* **117**(4), 889–901.

Ackermann, H. W. (2007). 5500 Phages examined in the electron microscope. *Arch. Virol.* **152**(2), 227–243.

Allinquant, B., *et al.* (1995). Downregulation of amyloid precursor protein inhibits neurite outgrowth in vitro. *J. Cell Biol.* **128**(5), 919–927.

Arap, W., Pasqualini, R., and Ruoslahti, E. (1998). Cancer treatment by targeted drug delivery to tumor vasculature in a mouse model. *Science* **279**(5349), 377–380.

Arap, W., *et al.* (2002). Steps toward mapping the human vasculature by phage display. *Nat. Med.* **8**(2), 121–127.

Chiu, Y. L., *et al.* (2004). Visualizing a correlation between siRNA localization, cellular uptake, and RNAi in living cells. *Chem. Biol.* **11**(8), 1165–1175.

Choi, M., *et al.* (2003). Inhibition of NF-kappaB by a TAT-NEMO-binding domain peptide accelerates constitutive apoptosis and abrogates LPS-delayed neutrophil apoptosis. *Blood* **102**(6), 2259–2267.

Dasgupta, S., *et al.* (2004). Antineuroinflammatory effect of NF-kappaB essential modifier-binding domain peptides in the adoptive transfer model of experimental allergic encephalomyelitis. *J. Immunol.* **173**(2), 1344–1354.

Dave, S. H., *et al.* (2007). Amelioration of chronic murine colitis by peptide-mediated transduction of the IkappaB kinase inhibitor NEMO binding domain peptide. *J. Immunol.* **179**(11), 7852–7859.

Derossi, D., *et al.* (1994). The third helix of the Antennapedia homeodomain translocates through biological membranes. *J. Biol. Chem.* **269**(14), 10444–10450.

Eguchi, A., *et al.* (2001). Protein transduction domain of HIV-1 Tat protein promotes efficient delivery of DNA into mammalian cells. *J. Biol. Chem.* **276**(28), 26204–26210.

Endoh, T., Sisido, M., and Ohtsuki, T. (2008). Cellular siRNA delivery mediated by a cell-permeant RNA-binding protein and photoinduced RNA interference. *Bioconjug. Chem.* **19**(5), 1017–1024.

Fawell, S., *et al.* (1994). Tat-mediated delivery of heterologous proteins into cells. *Proc. Natl. Acad. Sci. USA* **91**(2), 664–668.

Frankel, A. D., and Pabo, C. O. (1988). Cellular uptake of the tat protein from human immunodeficiency virus. *Cell* **55**(6), 1189–1193.

Futaki, S., *et al.* (2001). Arginine-rich peptides. An abundant source of membrane-permeable peptides having potential as carriers for intracellular protein delivery. *J. Biol. Chem.* **276**(8), 5836–5840.

Ghosh, A., *et al.* (2007). Selective inhibition of NF-kappaB activation prevents dopaminergic neuronal loss in a mouse model of Parkinson's disease. *Proc. Natl. Acad. Sci. USA* **104**(47), 18754–18759.

Green, M., and Loewenstein, P. M. (1988). Autonomous functional domains of chemically synthesized human immunodeficiency virus tat trans-activator protein. *Cell* **55**(6), 1179–1188.

Hayden, M. S., and Ghosh, S. (2008). Shared principles in NF-kappaB signaling. *Cell* **132**(3), 344–362.

Joliot, A. H., *et al.* (1991). alpha-2,8-Polysialic acid is the neuronal surface receptor of antennapedia homeobox peptide. *New Biol.* **3**(11), 1121–1134.

Josephson, L., *et al.* (1999). High-efficiency intracellular magnetic labeling with novel superparamagnetic-Tat peptide conjugates. *Bioconjug. Chem.* **10**(2), 186–191.

Kelly, K. A., Waterman, P., and Weissleder, R. (2006). In vivo imaging of molecularly targeted phage. *Neoplasia* **8**(12), 1011–1018.

Kolonin, M. G., *et al.* (2004). Reversal of obesity by targeted ablation of adipose tissue. *Nat. Med.* **10**(6), 625–632.

Kumar, P., *et al.* (2007). Transvascular delivery of small interfering RNA to the central nervous system. *Nature* **448**(7149), 39–43.

Lewin, M., *et al.* (2000). Tat peptide-derivatized magnetic nanoparticles allow in vivo tracking and recovery of progenitor cells. *Nat. Biotechnol.* **18**(4), 410–414.

Lundberg, P., *et al.* (2007). Delivery of short interfering RNA using endosomolytic cell-penetrating peptides. *FASEB J.* **21**(11), 2664–2671.

Mai, J. C., *et al.* (2001). A proapoptotic peptide for the treatment of solid tumors. *Cancer Res.* **61**(21), 7709–7712.

Mai, J. C., *et al.* (2002). Efficiency of protein transduction is cell type-dependent and is enhanced by dextran sulfate. *J. Biol. Chem.* **277**(33), 30208–30218.

May, M. J., *et al.* (2000). Selective inhibition of NF-kappaB activation by a peptide that blocks the interaction of NEMO with the IkappaB kinase complex. *Science* **289**(5484), 1550–1554.

Mi, Z., *et al.* (2003). Identification of a synovial fibroblast-specific protein transduction domain for delivery of apoptotic agents to hyperplastic synovium. *Mol. Ther.* **8**(2), 295–305.

Molenaar, T. J., *et al.* (2002). Uptake and processing of modified bacteriophage M13 in mice: Implications for phage display. *Virology* **293**(1), 182–191.

Morris, M. C., *et al.* (1997). A new peptide vector for efficient delivery of oligonucleotides into mammalian cells. *Nucleic Acids Res.* **25**(14), 2730–2736.

Moschos, S. A., *et al.* (2007). Lung delivery studies using siRNA conjugated to TAT(48-60) and penetratin reveal peptide induced reduction in gene expression and induction of innate immunity. *Bioconjug. Chem.* **18**(5), 1450–1459.

Muratovska, A., and Eccles, M. R. (2004). Conjugate for efficient delivery of short interfering RNA (siRNA) into mammalian cells. *FEBS Lett.* **558**(1–3), 63–68.

Pasqualini, R., McDonald, D. M., and Arap, W. (2001). Vascular targeting and antigen presentation. *Nat. Immunol.* **2**(7), 567–568.

Pooga, M., *et al.* (1998). Cell penetration by transportan. *FASEB J.* **12**(1), 67–77.

Ribeiro, M. M., *et al.* (2003). Heme oxygenase-1 fused to a TAT peptide transduces and protects pancreatic beta-cells. *Biochem. Biophys. Res. Commun.* **305**(4), 876–881.

Rothbard, J. B., *et al.* (2000). Conjugation of arginine oligomers to cyclosporin A facilitates topical delivery and inhibition of inflammation. *Nat. Med.* **6**(11), 1253–1257.

Schwarze, S. R., *et al.* (1999). In vivo protein transduction: Delivery of a biologically active protein into the mouse. *Science* **285**(5433), 1569–1572.

Scott, J. K., and Smith, G. P. (1990). Searching for peptide ligands with an epitope library. *Science* **249**(4967), 386–390.

Smith, G. P. (1985). Filamentous fusion phage: Novel expression vectors that display cloned antigens on the virion surface. *Science* **228**(4705), 1315–1317.

Strickland, I., and Ghosh, S. (2006). Use of cell permeable NBD peptides for suppression of inflammation. *Ann. Rheum. Dis.* **65**(Suppl 3), iii75–iii82.

Tilstra, J., *et al.* (2007). Protein transduction: Identification, characterization and optimization. *Biochem. Soc. Trans.* **35**(Pt 4), 811–815.

Torchilin, V. P., *et al.* (2001). TAT peptide on the surface of liposomes affords their efficient intracellular delivery even at low temperature and in the presence of metabolic inhibitors. *Proc. Natl. Acad. Sci. USA* **98**(15), 8786–8791.

Zhang, L., Hoffman, J. A., and Ruoslahti, E. (2005). Molecular profiling of heart endothelial cells. *Circulation* **112**(11), 1601–1611.

5

GRP78 Signaling Hub: A Receptor for Targeted Tumor Therapy

Masanori Sato, Virginia J. Yao, Wadih Arap, and Renata Pasqualini
David H. Koch Center, The University of Texas M.D. Anderson Cancer Center, Houston, Texas, USA

Advances in Genetics, Vol. 69
Copyright 2010, Elsevier Inc. All rights reserved.
0065-2660/10 $35.00
DOI: 10.1016/S0065-2660(10)69006-2

ABSTRACT

Glucose-regulated protein 78 (GRP78) is a potential receptor for targeting
therapy in cancer and chronic vascular disease due to its overexpression at the
cell surface in tumor cells and in atherosclerotic lesions. Presence of the GRP78
autoantibody in cancer patient sera is generally associated with poor prognosis
since it signals a prosurvival mechanism in response to cellular stress. Association of GRP78 with various binding partners involves coordination of multiple
signaling pathways that result in either cell survival or cell death. Binding of
activated α2-macroglobulin to cell-surface GRP78 activates Akt to suppress
apoptotic pathways through multiple downstream effectors, and concomitantly
upregulates NF-κB and induces the unfolded protein response (UPR) so that cell
proliferation prevails. Interaction of GRP78 with cell-surface T-cadherin promotes endothelial cell survival. Association of oncogenic Cripto with GRP78
nullifies TGF-β superfamily-dependent signaling through Smad2/3 to promote
cell proliferation. In contrast, association of GRP78 with the plasminogen
kringle 5 domain or extracellular Par-4 promotes apoptosis. Interaction of
GRP78 with microplasminogen induces the UPR while association with tissue
factor inhibits procoagulant activity. The diverse and multiple binding proteins of
GRP78 and their equally diverse functional outcomes reflect the regulatory
cellular functions that GRP78 orchestrates. Several GRP78 targeting peptides
have been isolated from different tumors and they show remarkable tumor
specificity. Conjugation of GRP78-targeting peptides to an apoptosis-inducing
peptide suppresses tumor growth in tumor xenografts, thereby demonstrating that
GRP78 is a viable target by which clinical cancer therapies can be successfully
developed as well as its potential utility in treating vascular disease.   © 2010, Elsevier Inc.

ABBREVIATIONS

A1-I3, α1 inhibitor III; ApoE, apolipoprotein E; ASK1, apoptosis signal-regulating kinase 1; ATF4, 6, activating transcription factor-4 or 6; Bad, Bcl-2-asso-
ciated death promoter; BiP, immunoglobulin heavy-chain binding protein;
CHOP, CCAAT/enhancer-binding protein homologous protein; CIA, collagen-induced arthritis; Cripto, teratocarcinoma-derived growth factor 1; EGF,
epidermal growth factor; eIF2α, eukaryotic translation initiation factor-2α; ER,
endoplasmic reticulum; FADD, fas-associated protein with death domain;
FOXO1, forkhead box O1; GADD34, growth arrest and DNA damage protein;
GADD45β, growth arrest and DNA-damage-inducible β; GRP78, glucose-regulated protein 78; GSK3β, glycogen synthase kinase-3β; IL, interleukin; IRE1-α,
inositol-requiring protein 1α; IKKα, IκB kinase α; JNK, c-Jun N-terminal
kinase; K5, Kringle 5; MAPK, mitogen-activated protein kinase; MIF,

macrophage migration inhibitory factor; MKK7, mitogen activated protein kinase kinase 7; MTJ-1, murine tumor cell DnaJ-like protein 1; NF-κB, nuclear factor κ beta; NLS1 and 2, nuclear localization signal 1 and 2; Par-4, prostate apoptosis response-4; PERK, PKR-like endoplasmic reticulum kinase; PI3K, phosphatidylinositol 3-kinase; PIN, prostatic intraepithelial neoplasia; PKA, protein kinase A; PKCζ, protein kinase C ζ isotype; RA, rheumatoid arthritis; SAC, selective for apoptosis in cancer cells; TF, tissue factor; TGFRI and RII, TGF-β type I and II receptor; TGF-β, transforming growth factor-β; Th1, 2, T-helper type 1, 2; TRAIL, TNF-related apoptosis-inducing ligand; UPR, unfolded protein response; VDAC, voltage-dependent anion channel; XIAP, x-linked inhibitor of apoptosis protein; α2M*, activated α2-macroglobulin (also known as low-density lipoprotein receptor-related protein 1).

I. INTRODUCTION

Current antiangiogenic cancer drugs comprise of either small molecule inhibitors or monoclonal antibodies (Heath and Bicknell, 2009). Similar to systemic cytotoxic drugs, multitargeted, small molecule antiangiogenic drugs cause toxicities in patients by interfering with concomitant growth factor signaling pathways that are involved in hematopoiesis, myelopoiesis, and endothelial cell survival and may induce drug resistance (Bergers and Hanahan, 2008; Verheul and Pinedo, 2007). Monoclonal antibodies such as bevacizamab or trastuzumab act extracellularly by binding either to a soluble growth factor ligand or membrane-bound growth factor receptor tyrosine kinase (Ferrara *et al.*, 2005; Goldenberg, 1999). The abundance of an angiogenic tumor vasculature that supports tumor growth suggests that accessible tumor-specific cell-surface proteins on the tumor vasculature or on tumor cells may be utilized as sites for tumor-specific delivery of anticancer drugs. Targeted therapeutic anticancer drugs are designed to be safer than nontargeted drugs since they are administered at lower doses while attaining a higher effective microconcentration at the tumor site(s) thereby reducing systemic toxic side effects.

Glucose-regulated protein 78 (GRP78), also known as immunoglobulin heavy-chain binding protein (BiP), is an endoplasmic reticulum (ER) resident molecular chaperone protein and a member of the heat shock protein 70 family (Li and Lee, 2006). GRP78 plays a central role in the unfolded protein response (UPR), an evolutionarily conserved mechanism in which survival or apoptotic pathways are activated in response to ER stress to physiologically reestablish normal function by translation repression, reduction of intermediate protein aggregates by increased folding, apoptosis of improperly folded proteins, and regulation of intracellular Ca^{2+} (Ma and Hendershot, 2004). Cell-surface GRP78 is detected by surface biotinylation in many types of cancer cells *in vivo*

(Delpino and Castelli, 2002; Triantafilou *et al.*, 2001) and is not present on noncancerous cells (Arap *et al.*, 2004). Expression of GRP78 on the surface of human cancer cells is associated with tumorgenesis, tumor progression, angiogenesis, and metastasis, thereby implicating its utility as a prognostic marker and a potential anticancer therapeutic target (Lee, 2007).

Evidence for GRP78 as a molecular tumor marker has been shown in a number of studies. Genetic reduction of GRP78 expression in $Grp78^{+/-}$ transgenic MMTVPyVT mice increases the latency period of spontaneous mammary tumors, decreases tumor growth rate, size, and microvessel density compared to transgenic $Grp78^{+/+}$ littermates (Dong *et al.*, 2008). Moreover, cultured tumor cells from $Grp78^{+/-}$/PyVT mice exhibit slower proliferation and increased apoptosis compared to cells isolated from $Grp78^{+/+}$/PyVT mice. In human gastric cancer, the Sp1 transcription factor is associated with advanced disease, poor prognosis, and correlates significantly with GRP78 expression (Zhang *et al.*, 2006). siRNA silencing of GRP78 expression inhibited both migration of NCI-N87 and SK-GT5 gastric tumor cells *in vitro* and xenograft tumor growth *in vivo*. PTEN-inactivated mice containing 1 wild-type $Grp78$ allele show invasive adenocarcinoma in the anterior prostate lobe, whereas adenocarcinoma is present in all three prostate lobes in $Grp78^{+/+}$ mice (Fu *et al.*, 2007). These studies also show that postnatal homozygous inactivation of GRP78 inhibits prostate cancer development initiated by the loss of the tumor suppressor PTEN. Studies of human glioblastoma, gastric, lung, prostate, and breast tumors show 50–75% of these tumors have enhanced levels of GRP78 that correlate with high pathologic grade and poor patient prognosis (Lee *et al.*, 2006, 2008; Pootrakul *et al.*, 2006; Wang *et al.*, 2005; Zhang *et al.*, 2006; Zheng *et al.*, 2008). GRP78 mRNA is detected at all tumor stages in esophageal cancer with higher expression in local early tumor growth and advanced tumor stages (Langer *et al.*, 2008). In contrast to most studies, high GRP78 expression in neuroblastoma and lung cancer is associated with a favorable prognosis (Hsu *et al.*, 2005; Uramoto *et al.*, 2005). These disparities possibly reflect the physiological functional complexity of GRP78 in the tumor microenvironment in conjunction with its role in the UPR cascade. Indeed, silencing GRP78 induces CCAAT/enhancer-binding protein homologous protein (CHOP), a proapoptotic effector protein in the UPR, and activates caspase-7 in temozolomide-treated malignant glioblastoma cells (Pyrko *et al.*, 2007). High expression levels of cell-surface GRP78 expression in human tumors are predominantly associated with cell proliferation, cell survival, angiogenesis, and metastasis (Lee, 2007).

Expression of GRP78 on cancer cells is potentiated under hypoxic conditions in which tumor angiogenesis, glucose metabolism, and invasion are supported by activating the hypoxia inducible factor 1 signaling pathway (Gonzalez-Gronow *et al.*, 2007; Semenza, 2009). Enhanced levels of GRP78 are localized to atherosclerotic lesions of aortic sections from apolipoprotein E

(ApoE) knockout mice as well as resected human arterial atherosclerotic lesions (Bhattacharjee et al., 2005; Liu et al., 2003). Herein, we review the multiple functional roles of cell-surface GRP78 through its association with diverse proteins and integration of signaling pathways to coordinate cell survival or cell death. Cell-surface GRP78 overexpression in human diseases such as cancer and atherosclerosis advocates its utility as a physiologically accessible and clinically relevant disease target. We propose GRP78 peptide ligands may be developed for targeted clinical therapies to treat human cancer and cardiovascular disease.

II. GRP78 ASSOCIATION WITH ACTIVATED α2-MACROGLOBULIN PROMOTES CELL PROLIFERATION

Cell-surface GRP78 associates with MHC class Ia, is a receptor for the coxsackie A9 and Dengue viruses, and functions as the signaling receptor upon binding to the activated form of the plasma proteinase inhibitor, α2-macroglogulin (α2M*) (Misra et al., 2009). Binding of GRP78 to α2M* in 1-LN human prostate tumor cells leads to cell proliferation (Misra et al., 2006). In the ER, the Dna-J murine homologue, MTJ-1, partners with GRP78 to maintain properly folded proteins (Chevalier et al., 2000). MTJ-1 translocates and anchors GRP78 to the cell surface in murine macrophages; silencing MTJ-1 expression at the RNA level attenuates GRP78 cell-surface localization and abolishes α2M* signaling (Misra et al., 2005). Association of cell-surface GRP78 with α2M* on 1-LN prostate tumor cells induces Akt phosphorylation that is PI3K-dependent (Misra et al., 2006). Active Akt phosphorylates downstream effectors such as FOXO1, Bad, GSK3β, and IKKα. Phosphorylated FOXO1, Bad, GSK3β, and IKKα promote cell proliferation either by inactivating apoptotic pathways or upregulating activated NF-κB. α2M* binding to GRP78 increases the expression of ER transmembrane genes such as PKR-like endoplasmic reticulum kinase (PERK), inositol-requiring protein 1 (IRE1-α), and activating transcription factor 6 (ATF6). The UPR transducers, PERK, IRE1-α, and ATF6 are normally inactive and bound to GRP78 in the ER. Under ER stress, PERK phosphorylates eukaryotic translation initiation factor-2α (eIF2α) thereby decreasing protein translation; translation arrest stimulates synthesis of ATF4, which induces growth arrest and DNA damage protein (GADD34) to dephosphorylate eIF2α and restores protein synthesis when ER stress is alleviated. Upregulation of NF-κB augments inactivation of mitogen-activated protein kinase kinase 7 (MKK7) through its binding to increased levels of growth arrest and DNA-damage-inducible β (GADD45β), thereby preventing JNK-mediated apoptosis. In addition, inactivation of apoptosis signal-regulating kinase (ASK1) by active Akt attenuates downstream JNK-mediated apoptosis. Akt phosphorylation of x-linked inhibitor of apoptosis protein (XIAP) prevents cell death by inhibiting caspase proteins. Thus, GRP78 binding to α2M* in 1-LN prostate tumor cells promotes cell proliferation by coordinating signals between mitogenic pathways in conjunction with the UPR cascade while inhibiting signaling pathways that lead to apoptosis.

Urothelial cells express GRP78, following substance P-induced bladder inflammation in rats (Vera et al., 2009). These studies established that urothelial cells release the proinflammatory cytokine, macrophage migration inhibitory factor (MIF) during inflammation. MIF complexes with A1-inhibitor 3 (A1-I3), an acute-phase protein of the α2 macroglobulin family. Binding of either MIF or the MIF–A1-I3 complex to cell-surface CD74 or GRP78, respectively, initiates the proinflammatory cascade. Intraluminal antibody inhibition of MIF, CD74, or GRP78 effectively reduces or prevents substance P-mediated inflammation.

III. GRP78 AUTOANTIBODY PREDICTS POOR PROGNOSIS IN TUMORS BUT SHOWS THERAPEUTIC EFFECTS IN RHEUMATOID ARTHRITIS

We previously reported 52% of patients with androgen-independent metastatic prostate cancer tested positive for the GRP78 autoantibody and its presence correlates with poor patient prognosis (Arap et al., 2004; Mintz et al., 2003). The GRP78 autoantibody has also been identified in the serum of ovarian cancer and rheumatoid arthritis (RA) patients (Bodman-Smith et al., 2004; Chinni et al., 1997; Taylor et al., 2009). The GRP78 autoantibody isolated from prostate cancer patients binds to Leu98-Leu115, which corresponds to the α2M* binding site (Gonzalez-Gronow et al., 2006). Similar to its association with α2M*, binding of GRP78 to the GRP78 autoantibody promotes cell survival and proliferation. In contrast, C-terminal anti-GRP78 antibodies act as receptor antagonists by blocking autophosphorylation and activation of GRP78 (Misra et al., 2009). Furthermore, treatment of 1-LN or DU145 prostate cancer cells, and A375 melanoma cells with C-terminal GRP78 antibodies results in DNA synthesis inhibition, upregulation of the p53 tumor suppressor, and apoptosis.

GRP78 may be a putative autoantigen in RA sera since the presence of GRP78 autoantibodies in the synovial fluid of RA patients is unique (Corrigall et al., 2001). Indeed, the frequency of RA patients with anti-GRP78 IgGs is reasonably high (63% of 400 patients), with no detectable expression levels in healthy subjects (Blass et al., 2001). GRP78 stimulates proliferating IL-10-producing CD8$^+$ T cells from normal patients (Bodman-Smith et al., 2003). The immune response to GRP78 in RA patients shifts the T-helper 1/2 (Th1/Th2) balance toward Th2, which promotes antiinflammation. GRP78 preimmunization inhibits development of collagen-induced arthritis (CIA) or pristane-induced arthritis in HLA-DR1$^{+/+}$ transgenic mice and Lewis rats (Corrigall et al., 2001).

Subcutaneous or intravenous immunization of DBA/1 mice with GRP78 effectively suppresses CIA by inducing the secretion of high levels of Th2 cytokines, IL-4, IL-5, and IL-10 by splenocytes and lymph node cells in the absence of adjuvant (Brownlie *et al.*, 2006). Cells collected from the spleen and lymph nodes of DBA/1 mice immunized with GRP78 showed a 2-fold proliferative response in cultures of splenocytes and lymph node cells compared to naïve mice or mice immunized with BSA. More importantly, transfer of lymph node cells and splenocytes from HLA-DR1$^{+/+}$ mice treated intravenously with 10 μg GRP78 to naïve HLA-DR1$^{+/+}$ mice immunized with autologous or homologous type II collagen effectively suppresses CIA. Taken together, preimmunization with GRP78 reveals novel immunomodulatory properties for potential therapeutic applications in RA patients.

IV. GRP78 PROMOTES ENDOTHELIAL CELL SURVIVAL IN ASSOCIATION WITH T-CADHERIN

Cell-surface GRP78 colocalizes with T-cadherin in human umbilical vein endothelial cells (HUVECs) (Philippova *et al.*, 2008). Unlike classical cadherins, T-cadherin lacks the cytosolic and transmembrane domains but retains the extracellular domain and binds to the plasma membrane via a glycosylphosphatidylinositol (GPI) anchor. It is expressed abundantly on the luminal surface of endothelial cells but not at sites of cell–cell contact; increased expression is observed in pathological conditions such as atherosclerotic lesions of the human aorta and in endothelial cells of the tumor vasculature (Ivanov *et al.*, 2004; Wyder *et al.*, 2000). Upregulated T-cadherin on human microvascular endothelial cells inactivates proapoptotic PERK in the UPR cascade, thereby protecting endothelial cells from ER-stress-induced apoptosis (Kyriakakis *et al.*, 2010). Overexpression of T-cadherin in HUVECs mediates cell survival in a GRP78-dependent fashion by increasing phospho-Akt and phospho-GSK3β and decreasing caspase-3 levels; N-terminal GRP78 antibodies abrogates Akt and GSK3β activation (Philippova *et al.*, 2008). The prosurvival effects of GRP78 in the UPR cascade and its association with overexpressed T-cadherin in the cardiovasculature and with cardiomyocytes suggest that GRP78 plays a central role in vascular tissue remodeling and stress.

V. CRIPTO MITOGENIC SIGNALING IS GRP78-DEPENDENT

Cell-surface binding of GRP78 to the GPI-anchored oncogene Cripto (Cripto-1, teratocarcinoma-derived growth factor 1) inhibits transforming growth factor-β (TGF-β) signaling and promotes cell proliferation (Shani *et al.*, 2008). Cripto is a member of the EGF–CFC protein family that functions as coreceptors of TGF-β ligands, such as Nodal, during embryonic development (Strizzi *et al.*, 2005). Cripto is normally absent in adult tissues but expressed at high levels in human tumors and associated with cell proliferation, migration, invasion, tumor angiogenesis, and epithelial-to-mesenchymal transition via activation of MAPK/ERK, PI3K/Akt, and inhibition of activin signaling. Oncogenic Cripto binds to TGF-β via its EGF domain to form a Cripto–TGF-β–TβRII complex that competes against TβRI binding to the TGF-β–TβRII complex. The Cripto–TGF-β –TβRII complex attenuates cytostatic TGF-β-mediated downstream Smad2 phosphorylation under anchorage-dependent and -independent conditions (Gray *et al.*, 2006).

The multiple mitogenic effects of Cripto that promote cell growth are transduced through its association with GRP78 at the cell surface (Shani *et al.*, 2008). The CFC domain of Cripto binds near the N-terminal domain of GRP78. Cripto-mediated activation of MAPK/PI3K pathways, downregulation of E-cadherin expression, and loss of cell adhesion in NCCIT embryonic carcinoma cells or MCF-10A cells are GRP78-dependent (Kelber *et al.*, 2009). The GRP78/ Cripto complex enhances cell proliferation through inhibition of Nodal or activin-A. Cripto or GRP878 knockdown leads to cell proliferation via inhibition of the antiproliferative activin-A cytokine, whereas the presence of the Cripto/GRP78 complex is necessary to attenuate growth inhibition by Nodal-dependent Smad2 phosphorylation.

VI. INTERACTION OF GRP78 WITH PLASMINOGEN KRINGLE 5 AND MICROPLASMINOGEN LEADS TO APOPTOSIS OR THE UPR

Proteolytic digestion of plasminogen produces potent angiogenic inhibitors such as angiostatin (kringles 1–4) and kringle 5 (Soff, 2000). Kringle domains consist of approximately 80 amino acid residues with three disulfide bonds. The recombinant Kringle 5 (rK5) domain promotes migration and induces apoptosis of proliferating human microvascular endothelial cells (HMEVC) stimulated with bFGF, aFGF, VEGF, PDGF, TGF-β1, IL-8, or HGF in a dose-dependent manner (Davidson *et al.*, 2005). Studies using the linear PRKLYDY peptide sequence within the K5 lysine-binding site identified GRP78 as the endothelial cell-surface binding protein. rK5 binding to GRP78 is competitively inhibited by the N-terminal polyclonal anti-GRP78 antibody in a dose-dependent fashion; anti-GRP78 blocks rK5-mediated proliferation, migration, and apoptosis of growth factor-stimulated HMEVC. rK5-mediated apoptosis of hypoxic HT1080 fibrosarcoma tumor cells correlates with stress-induced GRP78 expression and caspase-7.

Prior irradiation increases the sensitivity of brain microvessel endothelial cells (MeVC) to rK5-induced dose- and time-dependent apoptosis by 500-fold (McFarland *et al.*, 2009). rK5-induced apoptosis of irradiated or nonirradiated MvEC is GRP78-dependent, requires binding to α2M* and internalization, and activates p38MAPK. Although inactivation of p38MAPK downstream effectors such as ERK, MEK, or JNK does not inhibit activation of caspase-3, p38MAPK may be necessary to promote *grp78* transcription through ATF6 activation, thus reinforcing the apoptotic arm of the UPR (Li and Lee, 2006). The effects of rK5-induced apoptosis are proposed to be tumor specific and limited to MeVC in brain tumors and glioblastoma cells that express GRP78.

Although K5 binds to both GRP78 and the voltage-dependent anion channel (VDAC) on the cell surface of HUVEC and 1-LN prostate tumor cells under normoxic and hypoxic conditions, only VDAC mediates a Ca^{2+} signaling cascade (Gonzalez-Gronow *et al.*, 2007). K5 binding to VDAC in HUVEC elicits a small transient increase of intracellular Ca^{2+} from the extracellular milieu and decreases intracellular pH (Gonzalez-Gronow *et al.*, 2003, 2007). Intracellular acidification is consistent with early apoptotic events that induce hyperpolarization of the mitochondrial outer membrane. Silencing GRP78 expression on the surface of 1-LN cells by dsiRNA decreases VDAC expression (Gonzalez-Gronow *et al.*, 2007).

In addition to K5, another proteolytic product of plasminogen that contains a benzamidine binding site is microplasminogen (Gonzalez-Gronow *et al.*, 2007). Unlike K5, microplasminogen binds to the C-terminal region of GRP78. Stimulation of fura-2/AM preloaded 1-LN cells with the GRP78-binding microplasminogen peptide fragment, Ser759-Arg778, releases Ca^{2+} from intracellular stores and does not involve VDAC. Although K5 and microplasminogen bind to GRP78 near its N- and C-terminal regions, respectively, and both elicit increased intracellular Ca^{2+}, K5 binding elicits apoptosis while microplasminogen binding appears to be linked to the UPR cascade. These studies indicate that GRP78 regulates tumor cell viability and death by balancing external stresses inherent in the tumor microenvironment with internal factors through the UPR, particularly with alterations of intracellular Ca^{2+}, and its association with α2M*, K5, or microplasminogen.

VII. GRP78 ASSOCIATION WITH EXTRACELLULAR PAR-4 ACTIVATES THE EXTRINSIC APOPTOTIC PATHWAY

Prostate apoptosis response-4 (Par-4) gene was initially identified as an upregulated immediate early apoptotic gene in response to elevated intracellular Ca^{2+} in androgen-dependent and -independent AT-3 rat prostate tumor cells (Shrestha-Bhattarai and Rangnekar, 2010; Zhao and Rangnekar, 2008).

Human Par-4 protein was subsequently identified using a yeast two-hybrid screen as the binding partner of protein kinase Cζ (PKCζ) and Wilm's tumor protein, WT-1. Par-4 contains two putative nuclear localization sequences (NLS1 and 2) in the N-terminal region, a leucine zipper domain and nuclear export signal in the C-terminal domain. Akt phosphorylation of Par-4 and binding with the 14-3-3 chaperone protein in the cytosol prevent its translocation to the nucleus thereby promoting cell survival. In the presence of apoptotic stimuli, such as ionizing or UV radiation or increased intracellular Ca^{2+}, Par-4 reduces PKCζ activity by inducing a conformational change. Diminished PKCζ activity decreases FADD phosphorylation and IκB kinase activation. These events lead to the formation of the death inducing signaling complex and decreased NF-κB such that apoptosis ensues. The 58 amino acid core domain of Par-4 contains NLS2 and the Thr155 protein kinase A (PKA) phosphorylation site. This core domain is designated the selective apoptosis of cancer cells (SAC) domain and is 100% conserved in mammals and rodents. SAC overexpression leads to apoptosis via the extrinsic pathway.

Par-4 is a tumor suppressor and its downregulation is associated with renal-cell carcinoma, neuroblastoma, acute lymphoblastic leukemia, and chronic lymphocytic leukemia (Zhao and Rangnekar, 2008). Breast, gastric, pancreatic, and endometrial cancers are associated with *par-4* deletion or chromosomal instability, K-Ras point mutations, or gene silencing by DNA hypermethylation, respectively. In tumor cells, elevated PKA phosphorylates Par-4 at Thr155 and induces apoptosis upon nuclear translocation. SAC/B6C3F1 transgenic mice suppress the spontaneous development of hepatocarcinoma and lymphoma relative to Sac$^{-/-}$ control littermates. The SAC transgene inhibits TRAMP prostate tumor progression possibly by inducing apoptosis of PIN cells to delay progression to adenocarcinoma.

Par-4 is secreted by both normal and tumor cells by the conventional secretory pathway (Burikhanov *et al.*, 2009). Studies using PC-3 cells show extracellular Par-4 or SAC associates with cell-surface GRP78 near its N-terminus. Apoptosis of PC-3 cells is GRP78-dependent and occurs via ER stress and the FADD/caspase-8/caspase-3 pathway. Moreover, endogenous Par-4 expression is necessary for cell-surface GRP78 expression; silencing Par-4 expression does not reduce intracellular GRP78 levels. Pretreatment of PC-3 cells with the TRAIL proapoptotic protein or with ER-stress inducers, such as thapsigargin or tunicamycin, induces Par-4 secretion and expression of GRP78 and Par-4 at the plasma membrane.

Binding of the N-terminal domain of GRP78 with either K5 or secreted Par-4 elicits cell death. GRP78 binding with Par-4 elicits the extrinsic apoptotic pathway by activation of caspases-3 and -8 whereas GRP78/K5-mediated apoptosis involves the intrinsic pathway by activation of caspase-7 (Davidson *et al.*, 2005; Gonzalez-Gronow *et al.*, 2007).

VIII. GRP78 INHIBITS TISSUE FACTOR-MEDIATED PROCOAGULANT ACTIVITY

Cell-surface tissue factor (TF) is an integral membrane protein receptor that initiates the extrinsic coagulation cascade upon assembly with the serine protease, factors VII/VIIa to generate a functional TF:VIIa complex that rapidly activates the serine protease, factors X to Xa. The intrinsic and extrinsic clotting cascades converge at factor Xa, which complexes with activated factor V in the presence of Ca^{2+} and phospholipid to hydrolyze prothrombin to thrombin. Thrombin coordinates and sustains the procoagulation cascade by activating factor V, and factors VIII and XI of the intrinsic coagulation cascade. Thrombin cleaves fibrinogen to fibrin and activates factor XIII so that cross-linking of fibrin polymers by factor XIIIa forms a blood clot. Thrombin activation is regulated by the inhibitors, antithrombin III, heparin cofactor II, α_1-antitrypsin, and α_2 macroglobulin.

Early immunotherapy studies using a mouse melanoma or prostate xenograft tumor model indicated TF-mediated thrombosis is specifically targeted to tumor vascular endothelial cells or tumor cells to induce tumor infarction (Hu and Garen, 2001; Huang et al., 1997). Recent evidence, however, indicates TF-mediated signaling pathways that contribute to tumor growth, angiogenesis, and hematogenous metastases are deregulated in cancer and cancer-associated cells (Thomas et al., 2009). Studies using the murine Pan02 pancreatic tumor model show cancer cell-derived microparticles expressing TF and P-selectin glycoprotein ligand-1 circulate in the blood, accumulate at sites of injury in a P-selectin-dependent manner, and accelerate thrombus formation (Thomas et al., 2009). Although thrombosis is high in ovarian, prostate and lung adenocarcinomas, gastrointestinal carcinoma, and pancreatic cancer, abundant expression of TF in tumor, stromal, and endothelial cells within tumors and their permeable vasculature appears inconsistent with the general absence of thrombosis in other tumors (Hu and Garen, 2001; Ozawa et al., 2008; Pasqualini et al., 2002; Thomas et al., 2009). It has been suggested that endothelial cells, pericytes, stromal cells, as well as cancer cells, have developed mechanisms to attenuate thrombosis. Indeed, identification of GRP78 on microparticles shed from activated endothelial cells suggests GRP78 expression may regulate tumor hemorrhage and thrombosis (Banfi et al., 2005).

Studies using phage display technology to screen for peptide ligands that bind to endothelial surfaces of atherosclerotic lesions identified a peptide, EKP130, that binds GRP78 (Liu et al., 2003). Earlier studies show cholesterol accumulation causes ER stress in macrophages that leads to apoptosis, plaque rupture, thrombus formation, and vessel occlusion (Feng et al., 2003). Although ER stress/UPR activation occurs concurrently with atherosclerotic development in ApoE-deficient mice, apoptosis is not evident in early atherosclerotic lesions

(Zhou *et al.*, 2005). Taken together, these results indicate other signaling pathways may be involved that delay macrophage apoptosis until advanced atherosclerotic lesions form. GRP78 overexpression in T24/83 bladder cancer cells and cultured human aortic smooth muscle cells (HASMC) inhibits tissue factor-mediated procoagulant activity on the cell surface (Watson *et al.*, 2003). Cell-surface binding of TF to the C-terminal region of GRP78 was verified in adeno-viral-*grp78*-infected murine brain endothelial cells and macrophage cells (Bhattacharjee *et al.*, 2005; Pozza and Austin, 2005). GRP78 overexpression in HL-60, T24/83, or HASMC cells abrogates ionomycin or hydrogen perioxide activation of TF-mediated procoagulant activity (Bach and Moldow, 1997; Watson *et al.*, 2003). Since intracellular Ca^{2+} regulates anionic phospholipids on the outer plasma membrane, both of which are essential for TF procoagulant activity, another role by which GRP78 may attenuate thrombosis is to limit TF accessibility to cell-surface anionic phospholipids (Pozza and Austin, 2005).

These studies clearly indicate that tumor infarction by TF is not as straightforward as previously demonstrated since thrombosis in tumors is variable. This may reflect the influence and availability of GRP78 and Ca^{2+} to TF within the tumor microenviroment as well as unregulated signaling pathways that drive tumor growth, proliferation, angiogenesis, and metastases. Future investigation of the GRP78-TF regulatory axis in the vascular milieu will enhance our understanding of procoagulation in diseases such as cancer and vascular diseases such as myocardial and cerebral arterial thrombosis, so that appropriate clinical approaches can be designed.

IX. GRP78 TARGETING PEPTIDES INHIBIT TUMOR GROWTH

The GRP78-binding peptides, WIFPWIQL and WDLAWMFRLPVG, bind to endothelial cells that line the tumor vasculature *in vivo* (Arap *et al.*, 2004; Ozawa *et al.*, 2008). Conjugation of these peptides to the apoptosis-inducing sequence, $_D$(KLAKLAK)$_2$, abrogates tumor growth in mice without adverse side effects, whereas unconjugated WIFPWIQL, WDLAWMFRLPVG, or $_D$(KLAKLAK)$_2$ have no biological effect (Arap *et al.*, 2004). Despite ubiquitous expression of GRP78 in mammals, GRP78 peptide-presenting phage do not home to normal organs. Isolation of the CTVALPGGYVRVC peptide by screening a phage peptide library on human Me6652/4 melanoma cells identified its binding partner to be GRP78 (Kim *et al.*, 2006). Conjugation of CTVALPGGYVRVC with $_D$(KLAKLAK)$_2$ induces apoptosis in Me6652/4 and SJSA-1 osteosarcoma cells in a GRP78-dependent manner (Liu *et al.*, 2007). Since CTVALPGGYVRVC is homologous to plasminogen Ser759-Phe778, its interaction site with GRP78 is projected to be the C-terminal region (Gonzalez-Gronow *et al.*, 2007).

These studies show identification and specificity of different GRP78-targeting peptides from different tumors validates GRP78 as an ideal tumor target for clinical intervention.

Targeting GRP78 in tumors treated previously with antiangiogenic or chemotherapeutic drugs may eliminate the surviving tumor vasculature or cells in chemoresistant cancers (Lee *et al.*, 2006). Disruption of the tumor vasculature induces ER stress by glucose deprivation and hypoxia in the tumor microenvironment, which in turn increases GRP78 activity for the UPR and GRP78 expression (Sato *et al.*, 2007). Indeed, treatment of MDA-MB-435 tumor xenografts with contortrostatin, a disintegin, inhibits tumor-induced angiogenesis by blocking signaling pathways mediated by $\alpha_5\beta_1$, $\alpha_v\beta_3$, and $\alpha_v\beta_5$ integrins and induces transcriptional activation of GRP78 in surviving tumor cells (Dong *et al.*, 2005). Consistent with this result, treatment of human breast cancer MCF-7 cells with the topoisomerase II inhibitors, doxorubicin or etoposide (VP-16), produce resistant sublines that overexpress GRP78. Concurrent silencing of GRP78 expression and treatment with VP-16, significantly increases cell death of cultured MDA-MB-435 cells compared to control cells treated with VP-16 alone.

In addition to its therapeutic utility in cancer treatment, GRP78-targeting peptides show therapeutic promise to inhibit atherothrombosis or promote angiogenesis in ischemic tissues. Identification of a GRP78 binding peptide, CAPGPSKSC, by phage display on the endothelial surface of atherosclerotic lesions of ApoE knockout mice indicates increasing GRP78 expression or its association with TF may mitigate atherothrombotic disease (Liu *et al.*, 2003; Pozza and Austin, 2005). Another GRP78 binding peptide, YPHIDSLGHWRR, induces angiogenesis *in vitro* under hypoxic conditions by increasing endothelial cell proliferation, migration, and tube formation (Hardy *et al.*, 2008). Intramuscular injection of the YPHIDSLGHWRR peptide both restores blood perfusion and increases capillary density in a hind-limb ischemia mouse model.

X. CONCLUSIONS

We reviewed recent findings regarding cell-surface GRP78 association with various proteins. Binding of GRP78 via either its N- or C-terminal domains mediates multiple signaling pathways in both tumor and endothelial cells. Association to the N-terminal region of cell-surface GRP78 with α2M*, GRP78 autoantibody, T-cadherin, or Cripto results in cell proliferation and survival, whereas binding of GRP78 to extracellular Par-4 or K5 in association with VDAC results in apoptosis. GRP78 has been proposed as an autoantigen in RA patients, and studies in the CIA mouse model indicate a promising clinical application. Association of the C-terminal region of GRP78 with microplasminogen may involve the UPR pathway whereas GRP78 binding to TF inhibits procoagulation activity. Peptides that bind cell-surface GRP78 have been isolated from different tumors and atherosclerotic lesions and show potential for novel therapeutic strategies. For example, conjugation of GRP78 targeting peptides to $_D$(KLAKLAK)$_2$ specifically inhibits tumor growth, while another GRP78 binding peptide induces angiogenesis in peripheral ischemic tissues. The physiological accessibility of overexpressed cell-surface GRP78 and its pivotal role in determining cell survival or cell death in the tumor microenvironment favors it an ideal target receptor by which clinical therapies can be rationally designed to treat cancer in prospective clinical trials.

References

Arap, M. A., Lahdenranta, J., Mintz, P. J., Hajitou, A., Sarkis, A. S., Arap, W., and Pasqualini, R. (2004). Cell surface expression of the stress response chaperone GRP78 enables tumor targeting by circulating ligands. *Cancer Cell* **6**, 275–284.

Bach, R. R., and Moldow, C. F. (1997). Mechanism of tissue factor activation on HL-60 cells. *Blood* **89**, 3270–3276.

Banfi, C., Brioschi, M., Wait, R., Begum, S., Gianazza, E., Pirillo, A., Mussoni, L., and Tremoli, E. (2005). Proteome of endothelial cell-derived procoagulant microparticles. *Proteomics* **5**, 4443–4455.

Bergers, G., and Hanahan, D. (2008). Modes of resistance to anti-angiogenic therapy. *Nat. Rev. Cancer* **8**, 592–603.

Bhattacharjee, G., Ahamed, J., Pedersen, B., El-Sheikh, A., Mackman, N., Ruf, W., Liu, C., and Edgington, T. S. (2005). Regulation of tissue factor-mediated initiation of the coagulation cascade by cell surface grp78. *Arterioscler. Thromb. Vasc. Biol.* **25**, 1737–1743.

Blass, S., Union, A., Raymackers, J., Schumann, F., Ungethum, U., Muller-Steinbach, S., De Keyser, F., Engel, J. M., and Burmester, G. R. (2001). The stress protein BiP is overexpressed and is a major B and T cell target in rheumatoid arthritis. *Arthritis Rheum.* **44**, 761–771.

Bodman-Smith, M. D., Corrigall, V. M., Kemeny, D. M., and Panayi, G. S. (2003). BiP, a putative autoantigen in rheumatoid arthritis, stimulates IL-10-producing CD8-positive T cells from normal individuals. *Rheumatology (Oxford)* **42**, 637–644.

Bodman-Smith, M. D., Corrigall, V. M., Berglin, E., Cornell, H. R., Tzioufas, A. G., Mavragani, C. P., Chan, C., Rantapaa-Dahlqvist, S., and Panayi, G. S. (2004). Antibody response to the human stress protein BiP in rheumatoid arthritis. *Rheumatology (Oxford)* **43**, 1283–1287.

Brownlie, R. J., Myers, L. K., Wooley, P. H., Corrigall, V. M., Bodman-Smith, M. D., Panayi, G. S., and Thompson, S. J. (2006). Treatment of murine collagen-induced arthritis by the stress protein BiP via interleukin-4-producing regulatory T cells: A novel function for an ancient protein. *Arthritis Rheum.* **54**, 854–863.

Burikhanov, R., Zhao, Y., Goswami, A., Qiu, S., Schwarze, S. R., and Rangnekar, V. M. (2009). The tumor suppressor Par-4 activates an extrinsic pathway for apoptosis. *Cell* **138**, 377–388.

Chevalier, M., Rhee, H., Elguindi, E. C., and Blond, S. Y. (2000). Interaction of murine BiP/GRP78 with the DnaJ homologue MTJ1. *J. Biol. Chem.* **275**, 19620–19627.

Chinni, S. R., Falchetto, R., Gercel-Taylor, C., Shabanowitz, J., Hunt, D. F., and Taylor, D. D. (1997). Humoral immune responses to cathepsin D and glucose-regulated protein 78 in ovarian cancer patients. *Clin. Cancer Res.* **3**, 1557–1564.

Corrigall, V. M., Bodman-Smith, M. D., Fife, M. S., Canas, B., Myers, L. K., Wooley, P., Soh, C., Staines, N. A., Pappin, D. J., Berlo, S. E., et al. (2001). The human endoplasmic reticulum molecular chaperone BiP is an autoantigen for rheumatoid arthritis and prevents the induction of experimental arthritis. J. Immunol. 166, 1492–1498.

Davidson, D. J., Haskell, C., Majest, S., Kherzai, A., Egan, D. A., Walter, K. A., Schneider, A., Gubbins, E. F., Solomon, L., Chen, Z., et al. (2005). Kringle 5 of human plasminogen induces apoptosis of endothelial and tumor cells through surface-expressed glucose-regulated protein 78. Cancer Res. 65, 4663–4672.

Delpino, A., and Castelli, M. (2002). The 78 kDa glucose-regulated protein (GRP78/BIP) is expressed on the cell membrane, is released into cell culture medium and is also present in human peripheral circulation. Biosci. Rep. 22, 407–420.

Dong, D., Ko, B., Baumeister, P., Swenson, S., Costa, F., Markland, F., Stiles, C., Patterson, J. B., Bates, S. E., and Lee, A. S. (2005). Vascular targeting and antiangiogenesis agents induce drug resistance effector GRP78 within the tumor microenvironment. Cancer Res. 65, 5785–5791.

Dong, D., Ni, M., Li, J., Xiong, S., Ye, W., Virrey, J. J., Mao, C., Ye, R., Wang, M., Pen, L., et al. (2008). Critical role of the stress chaperone GRP78/BiP in tumor proliferation, survival, and tumor angiogenesis in transgene-induced mammary tumor development. Cancer Res. 68, 498–505.

Feng, B., Zhang, D., Kuriakose, G., Devlin, C. M., Kockx, M., and Tabas, I. (2003). Niemann-Pick C heterozygosity confers resistance to lesional necrosis and macrophage apoptosis in murine atherosclerosis. Proc. Natl. Acad. Sci. USA 100, 10423–10428.

Ferrara, N., Hillan, K. J., and Novotny, W. (2005). Bevacizumab (Avastin), a humanized anti-VEGF monoclonal antibody for cancer therapy. Biochem. Biophys. Res. Commun. 333, 328–335.

Fu, Y., Li, J., and Lee, A. S. (2007). GRP78/BiP inhibits endoplasmic reticulum BIK and protects human breast cancer cells against estrogen starvation-induced apoptosis. Cancer Res. 67, 3734–3740.

Goldenberg, M. M. (1999). Trastuzumab, a recombinant DNA-derived humanized monoclonal antibody, a novel agent for the treatment of metastatic breast cancer. Clin. Ther. 21, 309–318.

Gonzalez-Gronow, M., Kalfa, T., Johnson, C. E., Gawdi, G., and Pizzo, S. V. (2003). The voltage-dependent anion channel is a receptor for plasminogen kringle 5 on human endothelial cells. J. Biol. Chem. 278, 27312–27318.

Gonzalez-Gronow, M., Cuchacovich, M., Llanos, C., Urzua, C., Gawdi, G., and Pizzo, S. V. (2006). Prostate cancer cell proliferation in vitro is modulated by antibodies against glucose-regulated protein 78 isolated from patient serum. Cancer Res. 66, 11424–11431.

Gonzalez-Gronow, M., Kaczowka, S. J., Payne, S., Wang, F., Gawdi, G., and Pizzo, S. V. (2007). Plasminogen structural domains exhibit different functions when associated with cell surface GRP78 or the voltage-dependent anion channel. J. Biol. Chem. 282, 32811–32820.

Gray, P. C., Shani, G., Aung, K., Kelber, J., and Vale, W. (2006). Cripto binds transforming growth factor beta (TGF-beta) and inhibits TGF-beta signaling. Mol. Cell Biol. 26, 9268–9278.

Hardy, B., Battler, A., Weiss, C., Kudasi, O., and Raiter, A. (2008). Therapeutic angiogenesis of mouse hind limb ischemia by novel peptide activating GRP78 receptor on endothelial cells. Biochem. Pharmacol. 75, 891–899.

Heath, V. L., and Bicknell, R. (2009). Anticancer strategies involving the vasculature. Nat. Rev. Clin. Oncol. 6, 395–404.

Hsu, W. M., Hsieh, F. J., Jeng, Y. M., Kuo, M. L., Tsao, P. N., Lee, H., Lin, M. T., Lai, H. S., et al. (2005). GRP78 expression correlates with histologic differentiation and favorable prognosis in neuroblastic tumors. Int. J. Cancer 113, 920–927.

Hu, Z., and Garen, A. (2001). Targeting tissue factor on tumor vascular endothelial cells and tumor cells for immunotherapy in mouse models of prostatic cancer. Proc. Natl. Acad. Sci. USA 98, 12180–12185.

Huang, X., Molema, G., King, S., Watkins, L., Edgington, T. S., and Thorpe, P. E. (1997). Tumor infarction in mice by antibody-directed targeting of tissue factor to tumor vasculature. *Science* **275,** 547–550.

Ivanov, D., Philippova, M., Allenspach, R., Erne, P., and Resink, T. (2004). T-cadherin upregulation correlates with cell-cycle progression and promotes proliferation of vascular cells. *Cardiovasc. Res.* **64,** 132–143.

Kelber, J. A., Panopoulos, A. D., Shani, G., Booker, E. C., Belmonte, J. C., Vale, W. W., and Gray, P. C. (2009). Blockade of Cripto binding to cell surface GRP78 inhibits oncogenic Cripto signaling via MAPK/PI3K and Smad2/3 pathways. *Oncogene* **28,** 2324–2336.

Kim, Y., Lillo, A. M., Steiniger, S. C., Liu, Y., Ballatore, C., Anichini, A., Mortarini, R., Kaufmann, G. F., Zhou, B., Felding-Habermann, B., et al. (2006). Targeting heat shock proteins on cancer cells: Selection, characterization, and cell-penetrating properties of a peptidic GRP78 ligand. *Biochemistry* **45,** 9434–9444.

Kyriakakis, E., Philippova, M., Joshi, M. B., Pfaff, D., Bochkov, V., Afonyushkin, T., Erne, P., and Resink, T. J. (2010). T-cadherin attenuates the PERK branch of the unfolded protein response and protects vascular endothelial cells from endoplasmic reticulum stress-induced apoptosis. *Cell Signal.* **22,** 1308–1316.

Langer, R., Feith, M., Siewert, J. R., Wester, H. J., and Hoefler, H. (2008). Expression and clinical significance of glucose regulated proteins GRP78 (BiP) and GRP94 (GP96) in human adenocarcinomas of the esophagus. *BMC Cancer* **8,** 70.

Lee, A. S. (2007). GRP78 induction in cancer: Therapeutic and prognostic implications. *Cancer Res.* **67,** 3496–3499.

Lee, E., Nichols, P., Spicer, D., Groshen, S., Yu, M. C., and Lee, A. S. (2006). GRP78 as a novel predictor of responsiveness to chemotherapy in breast cancer. *Cancer Res.* **66,** 7849–7853.

Lee, H. K., Xiang, C., Cazacu, S., Finniss, S., Kazimirsky, G., Lemke, N., Lehman, N. L., Rempel, S. A., Mikkelsen, T., and Brodie, C. (2008). GRP78 is overexpressed in glioblastomas and regulates glioma cell growth and apoptosis. *Neuro Oncol.* **10,** 236–243.

Li, J., and Lee, A. S. (2006). Stress induction of GRP78/BiP and its role in cancer. *Curr. Mol. Med.* **6,** 45–54.

Liu, C., Bhattacharjee, G., Boisvert, W., Dilley, R., and Edgington, T. (2003). In vivo interrogation of the molecular display of atherosclerotic lesion surfaces. *Am. J. Pathol.* **163,** 1859–1871.

Liu, Y., Steiniger, S. C., Kim, Y., Kaufmann, G. F., Felding-Habermann, B., and Janda, K. D. (2007). Mechanistic studies of a peptidic GRP78 ligand for cancer cell-specific drug delivery. *Mol. Pharm.* **4,** 435–447.

Ma, Y., and Hendershot, L. M. (2004). The role of the unfolded protein response in tumour development: Friend or foe? *Nat. Rev. Cancer* **4,** 966–977.

McFarland, B. C., Stewart, J., Jr., Hamza, A., Nordal, R., Davidson, D. J., Henkin, J., and Gladson, C. L. (2009). Plasminogen kringle 5 induces apoptosis of brain microvessel endothelial cells: Sensitization by radiation and requirement for GRP78 and LRP1. *Cancer Res.* **69,** 5537–5545.

Mintz, P. J., Kim, J., Do, K. A., Wang, X., Zinner, R. G., Cristofanilli, M., Arap, M. A., Hong, W. K., Troncoso, P., Logothetis, C. J., et al. (2003). Fingerprinting the circulating repertoire of antibodies from cancer patients. *Nat. Biotechnol.* **21,** 57–63.

Misra, U. K., Gonzalez-Gronow, M., Gawdi, G., and Pizzo, S. V. (2005). The role of MTJ-1 in cell surface translocation of GRP78, a receptor for alpha 2-macroglobulin-dependent signaling. *J. Immunol.* **174,** 2092–2097.

Misra, U. K., Deedwania, R., and Pizzo, S. V. (2006). Activation and cross-talk between Akt, NF-kappaB, and unfolded protein response signaling in 1-LN prostate cancer cells consequent to ligation of cell surface-associated GRP78. *J. Biol. Chem.* **281,** 13694–13707.

Misra, U. K., Mowery, Y., Kaczowka, S., and Pizzo, S. V. (2009). Ligation of cancer cell surface GRP78 with antibodies directed against its COOH-terminal domain up-regulates p53 activity and promotes apoptosis. *Mol. Cancer Ther.* **8,** 1350–1362.

Ozawa, M. G., Zurita, A. J., Dias-Neto, E., Nunes, D. N., Sidman, R. L., Gelovani, J. G., Arap, W., and Pasqualini, R. (2008). Beyond receptor expression levels: The relevance of target accessibility in ligand-directed pharmacodelivery systems. *Trends Cardiovasc. Med.* **18,** 126–132.

Pasqualini, R., Arap, W., and McDonald, D. M. (2002). Probing the structural and molecular diversity of tumor vasculature. *Trends Mol. Med.* **8,** 563–571.

Philippova, M., Ivanov, D., Joshi, M. B., Kyriakakis, E., Rupp, K., Afonyushkin, T., Bochkov, V., Erne, P., and Resink, T. J. (2008). Identification of proteins associating with glycosylphosphatidylinositol- anchored T-cadherin on the surface of vascular endothelial cells: Role for Grp78/BiP in T-cadherin-dependent cell survival. *Mol. Cell Biol.* **28,** 4004–4017.

Pootrakul, L., Datar, R. H., Shi, S. R., Cai, J., Hawes, D., Groshen, S. G., Lee, A. S., and Cote, R. J. (2006). Expression of stress response protein Grp78 is associated with the development of castration-resistant prostate cancer. *Clin. Cancer Res.* **12,** 5987–5993.

Pozza, L. M., and Austin, R. C. (2005). Getting a GRP on tissue factor activation. *Arterioscler. Thromb. Vasc. Biol.* **25,** 1529–1531.

Pyrko, P., Schonthal, A. H., Hofman, F. M., Chen, T. C., and Lee, A. S. (2007). The unfolded protein response regulator GRP78/BiP as a novel target for increasing chemosensitivity in malignant gliomas. *Cancer Res.* **67,** 9809–9816.

Sato, M., Arap, W., and Pasqualini, R. (2007). Molecular targets on blood vessels for cancer therapies in clinical trials. *Oncology (Williston Park)* **21,** 1346–1352. (discussion 1354–1355, 1367, 1370 passim).

Semenza, G. L. (2009). Regulation of cancer cell metabolism by hypoxia-inducible factor 1. *Semin. Cancer Biol.* **19,** 12–16.

Shani, G., Fischer, W. H., Justice, N. J., Kelber, J. A., Vale, W., and Gray, P. C. (2008). GRP78 and Cripto form a complex at the cell surface and collaborate to inhibit transforming growth factor beta signaling and enhance cell growth. *Mol. Cell Biol.* **28,** 666–677.

Shrestha-Bhattarai, T., and Rangnekar, V. M. (2010). Cancer-selective apoptotic effects of extracellular and intracellular Par-4. *Oncogene* **29,** 3873–3880.

Soff, G. A. (2000). Angiostatin and angiostatin-related proteins. *Cancer Metastasis Rev.* **19,** 97–107.

Strizzi, L., Bianco, C., Normanno, N., and Salomon, D. (2005). Cripto-1: A multifunctional modulator during embryogenesis and oncogenesis. *Oncogene* **24,** 5731–5741.

Taylor, D. D., Gercel-Taylor, C., and Parker, L. P. (2009). Patient-derived tumor-reactive antibodies as diagnostic markers for ovarian cancer. *Gynecol. Oncol.* **115,** 112–120.

Thomas, G. M., Panicot-Dubois, L., Lacroix, R., Dignat-George, F., Lombardo, D., and Dubois, C. (2009). Cancer cell-derived microparticles bearing P-selectin glycoprotein ligand 1 accelerate thrombus formation in vivo. *J. Exp. Med.* **206,** 1913–1927.

Triantafilou, M., Fradelizi, D., and Triantafilou, K. (2001). Major histocompatibility class one molecule associates with glucose regulated protein (GRP) 78 on the cell surface. *Hum. Immunol.* **62,** 764–770.

Uramoto, H., Sugio, K., Oyama, T., Nakata, S., Ono, K., Yoshimastu, T., Morita, M., and Yasumoto, K. (2005). Expression of endoplasmic reticulum molecular chaperone Grp78 in human lung cancer and its clinical significance. *Lung Cancer* **49,** 55–62.

Vera, P. L., Wang, X., Bucala, R. J., and Meyer-Siegler, K. L. (2009). Intraluminal blockade of cell-surface CD74 and glucose regulated protein 78 prevents substance P-induced bladder inflammatory changes in the rat. *PLoS ONE* **4,** e5835.

Verheul, H. M., and Pinedo, H. M. (2007). Possible molecular mechanisms involved in the toxicity of angiogenesis inhibition. *Nat. Rev. Cancer* **7,** 475–485.

Wang, Q., He, Z., Zhang, J., Wang, Y., Wang, T., Tong, S., Wang, L., Wang, S., and Chen, Y. (2005). Overexpression of endoplasmic reticulum molecular chaperone GRP94 and GRP78 in human lung cancer tissues and its significance. *Cancer Detect. Prev.* **29,** 544–551.

Watson, L. M., Chan, A. K., Berry, L. R., Li, J., Sood, S. K., Dickhout, J. G., Xu, L., Werstuck, G. H., Bajzar, L., Klamut, H. J., *et al.* (2003). Overexpression of the 78-kDa glucose-regulated protein/immunoglobulin-binding protein (GRP78/BiP) inhibits tissue factor procoagulant activity. *J. Biol. Chem.* **278,** 17438–17447.

Wyder, L., Vitaliti, A., Schneider, H., Hebbard, L. W., Moritz, D. R., Wittmer, M., Ajmo, M., and Klemenz, R. (2000). Increased expression of H/T-cadherin in tumor-penetrating blood vessels. *Cancer Res.* **60,** 4682–4688.

Zhang, J., Jiang, Y., Jia, Z., Li, Q., Gong, W., Wang, L., Wei, D., Yao, J., Fang, S., and Xie, K. (2006). Association of elevated GRP78 expression with increased lymph node metastasis and poor prognosis in patients with gastric cancer. *Clin. Exp. Metastasis* **23,** 401–410.

Zhao, Y., and Rangnekar, V. M. (2008). Apoptosis and tumor resistance conferred by Par-4. *Cancer Biol. Ther.* **7,** 1867–1874.

Zheng, H. C., Takahashi, H., Li, X. H., Hara, T., Masuda, S., Guan, Y. F., and Takano, Y. (2008). Overexpression of GRP78 and GRP94 are markers for aggressive behavior and poor prognosis in gastric carcinomas. *Hum. Pathol.* **39,** 1042–1049.

Zhou, J., Lhotak, S., Hilditch, B. A., and Austin, R. C. (2005). Activation of the unfolded protein response occurs at all stages of atherosclerotic lesion development in apolipoprotein E-deficient mice. *Circulation* **111,** 1814–1821.

6 On the Synergistic Effects of Ligand-Mediated and Phage-Intrinsic Properties During *In Vivo* Selection

Wouter H. P. Driessen,[*,1] **Lawrence F. Bronk,**[*,1] **Julianna K. Edwards,**[*] **Bettina Proneth,**[*] **Glauco R. Souza,**[*,2] **Paolo Decuzzi,**[†,‡] **Renata Pasqualini,**[*,§] **and Wadih Arap**[*,§]

[*]David H. Koch Center, The University of Texas M.D. Anderson Cancer Center, Houston, Texas, USA

[†]Department of Nanomedicine and Biomedical Engineering, The University of Texas Health Science Center at Houston, Houston, Texas, USA

[‡]Center of Bio-/Nanotechnology and -/Engineering for Medicine, University of Magna Graecia, Catanzaro, Italy

[§]Department of Experimental Diagnostic Imaging, The University of Texas M.D. Anderson Cancer Center, Houston, Texas, USA

[1]These authors contributed equally to this work.
[2]Present address: Nano3D Biosciences, Inc., Houston, Texas, USA.

Advances in Genetics, Vol. 69
Copyright 2010, Elsevier Inc. All rights reserved.

0065-2660/10 $35.00
DOI: 10.1016/S0065-2660(10)69005-0

ABSTRACT

Phage display has been used as a powerful tool in the discovery and characterization of ligand–receptor complexes that can be utilized for therapeutic applications as well as to elucidate disease mechanisms. While the basic properties of phage itself have been well described, the behavior of phage in an *in vivo* setting is not as well understood due to the complexity of the system. Here, we take a dual approach in describing the biophysical mechanisms and properties that contribute to the efficacy of *in vivo* phage targeting. We begin by considering the interaction between phage and target by applying a kinetic model of ligand–receptor complexation and internalization. The multivalent display of peptides on the pIII capsid of phage is also discussed as an augmenting factor in the binding affinity of phage-displayed peptides to cellular targets accessible in a microenvironment of interest. Lastly, we examine the physical properties of the total phage particle that facilitate improved delivery and targeting *in vivo* compared to free peptides. © 2010, Elsevier Inc.

I. INTRODUCTION

Over the last years, drug developers have been channeling an increasing amount of effort and resources toward developing high-throughput screening methods for identifying potential drug targeting molecules based on the screening of ligand–receptor pairs (Dickson and Gagnon, 2004). Despite these efforts, the pharmaceutical industry has encountered a *"pipeline problem"*—a slowdown, instead of an acceleration, of innovative medical therapies reaching patients. This is reflected in the fact that the United States Food and Drug Administration (FDA) saw the number of new drug and biologic applications submitted to them for approval decline significantly and in response to this development, they started the "Critical Path Initiative." This initiative calls for a directed effort to modernize scientific tools and make use of the potential of bioinformatics (Lesko, 2007; Woodcock and Woosley, 2008). We believe that the *in vivo* screening of active compounds will inevitably result in the discovery of more advanced therapeutic candidates, resulting in significant savings of both time and resources compared to the traditional drug discovery process (see Fig. 6.1). One of the shortcomings

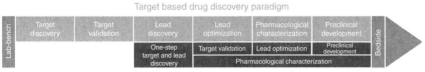

Figure 6.1. Target based versus *in vivo* drug discovery paradigms. The direct *in vivo* screening of phage-displayed peptide libraries allows for the one-step discovery of targets and lead compounds. Another advantage is that there is no bias toward the targeted receptor.

of *in vitro* target-based high-throughput screenings is that most empirical and theoretical approaches using combinatorial libraries are based on idealized conditions and local molecular interactions which are only snap shots of the actual dynamic and nonequilibrium cellular environment. There is limited value in such idealized approaches, which fail to account for existing nonequilibrium processes, such as membrane bound interactions, rates of receptor mediated internalization, global and local receptor/ligand concentrations, and other dynamic cellular processes. A few attempts have been made towards screening combinatorial libraries *in vivo*. Brown *et al.* (2004) used a mass spectrometry based approach to identify low molecular weight compounds based on tissue specificity and pharmacokinetics rather than defined activity. Although drug candidates were selected, the authors were not able to define the binding-partner for the selected ligand, complicating downstream lead optimization. Houghten *et al.* (2008) have recently explored the *in vivo* screening of mixture-based chemically diverse libraries as a new approach towards drug discovery. In this study, mixture-based libraries were designed, synthesized, and directly used in an *in vivo* mouse model of pain modulation. The most active sequence was subsequently identified using the iterative deconvolution approach, a step-by-step selection process to identify individual compounds. Hereby, the most active sequence is identified by the systematic reduction in the number of compounds in the most active mixture. A drawback of mixture-based chemically diverse libraries is the necessity of iterative deconvolution, which will require labor-intensive and costly resynthesis of library components (Houghten *et al.*, 2008). Biopanning using phage-displayed combinatorial peptide libraries is an *in vivo* screening technique that has been successfully and reproducibly used over the past decades to uncover accessible ligands without bias or preconceived notion towards the receptor being targeted (Pasqualini and Ruoslahti, 1996). Phage display involves genetically engineering bacteriophage so that a random combinatorial library of peptides can be expressed on their surface (Smith and Petrenko, 1997). Once a candidate target is identified, it can be isolated, purified, and cloned using biochemical molecular methods (Christianson *et al.*, 2007;

Kolonin *et al.*, 2006; Pasqualini *et al.*, 2010; Trepel *et al.*, 2008). As opposed to target-based screening, *in vivo* phage display screens directly in relevant (animal) models of disease or even directly in human patients (Arap *et al.*, 2002; Krag *et al.*, 2006). This approach will lead to the one-step identification of druggable targets and leads, thus saving valuable time and money in the drug discovery process and bringing promising novel drugs faster from the lab-bench to the bed-side (Fig. 6.1). Peptide-leads identified by disease-based screenings have high specificity and should have reduced off-target activities. Moreover, phage display derived leads are peptide-based, allowing for relatively simple and cheap peptide-chemistry in the lead-optimization process and for this drug discovery paradigm to be performed in academic laboratories.

To better understand the success of *in vivo* phage display, we present here some of the underlying mechanisms of phage circulation, binding, and internalization with a focus on peptide-mediated events and effects mediated by the phage itself, that is, intrinsic "phage-properties."

II. LIGAND–RECEPTOR MEDIATED EFFECTS

Phage and target cell interactions can be simplified into a traditional model of ligand–receptor binding. The multivalent display of peptides on the pIII minor capsid of the phage act as targeting ligands that facilitate specific binding and cellular internalization of the entire phage particle. These ligand–receptor mediated effects can be partially captured in a kinematic model in an attempt to explain the established efficacy of *in vivo* phage display applications.

A. Kinematic model for surface binding and internalization of phage

Here, we introduce a mathematical model for predicting the specific binding of phage-displayed peptides to receptor molecules expressed on the cell membrane and their receptor-mediated internalization. The formation of ligand–receptor bonds at the phage–cell interface can be described analogously to a chemical reaction having the form:

$$lL + rR \underset{k_r}{\overset{k_f}{\rightleftarrows}} cC \qquad (6.1)$$

where L, R, and C refer to the ligand molecules on the phage, the receptor molecules expressed on the cell membrane, and the complexes formed upon binding of the ligands with the receptors, respectively (Fig. 6.2). The binding process is characterized by forward k_f and reverse k_r reaction rates. In the case of phage-displayed peptides, l is the valency of the phage (i.e., the number of

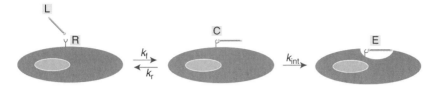

Figure 6.2. Phage mediated ligand–receptor interaction and internalization. Ligand (L) displayed on the pIII capsid of phage binds to its target receptor (R) with the rate k_f to form a ligand–receptor complex (C). Complex dissociates with the rate k_r. Following binding, the phage ligand–receptor complex is internalized (E) into the cell with the rate k_{int}.

peptides expressed at the phage's "active edge," usually pIII with a value between $l = 1$–5), whereas r and c equal unity (Barbas, 2001). The internalization of the complexes C (cell receptor bound phage) can again be described as a reaction occurring with a forward rate k_{int} so that:

$$C \xrightarrow{k_{int}} E \qquad (6.2)$$

where E refers to the internalized complex. Mass balance for the formation of the complex C requires that:

$$\frac{dC(t)}{dt} = +lk_f L(t)R(t) - k_r C(t) - k_{int} C(t) \qquad (6.3)$$

where $lk_f L(t)R(t)$ represents the increase in complex C concentration over time associated with the binding of ligand and receptor molecules, $k_r C(t)$ represents the reduction in complex C concentration over time associated with the "debinding" of ligand–receptor complexes, and $k_{int} C(t)$ represents the reduction in complex C concentration over time associated with the internalization of the formed ligand–receptor complexes. Under the assumption that the initial concentrations ($t = 0$) of phage (ligands) and cell receptors are large to be considered constant over time t, it follows that:

$$L(t) = L(0) \equiv L_o \quad \text{and} \quad R(t) = R(0) \equiv R_o \qquad (6.4)$$

Taken together with the initial condition $C(0) = C_o = 0$, Eq. (6.1) gives a closed form solution for $C(t)$ as:

$$C(t) = \frac{lk_f}{k_r + k_{int}} L_o R_o [1 - e^{-(k_r + k_{int})t}] \qquad (6.5)$$

Again, invoking mass balance, the concentrations of the internalized complex $E(t)$ and bound complex $C(t)$ are related through the differential equation:

$$\frac{dE(t)}{dt} = k_{int}C(t) \tag{6.6}$$

from which $E(t)$ can be readily derived after substituting in $C(t)$ from Eq. (6.5) and integrating with respect to time t, resulting in:

$$E(t) = \frac{lk_f k_{int}}{k_r + k_{int}} L_o R_o \left[t - \frac{1}{k_r + k_{int}} \left(1 - e^{-(k_r + k_{int})t}\right) \right] \tag{6.7}$$

Equations (6.5) and (6.7) provide explicit expressions for the bound ($C(t)$) and internalized ($E(t)$) phages over time. These equations can be conveniently rephrased by introducing the dimensionless parameters

$$\alpha = \frac{lk_f}{k_{int}}, \quad \beta = \frac{k_r}{k_{int}}, \quad \tau = k_{int}t \tag{6.8}$$

to derive

$$\frac{C(\tau)}{L_o R_o} = \frac{\alpha}{\beta + 1} \left[1 - e^{-(\beta+1)\tau}\right] \quad \text{and} \quad \frac{E(\tau)}{L_o R_o} = \frac{\alpha}{\beta + 1} \left[\tau - \frac{1 - e^{-(\beta+1)\tau}}{\beta + 1} \right] \tag{6.9}$$

From the analysis of Eq. (6.9), it can be observed that the concentration of bound $C(t)$ and internalized $E(t)$ phages increases as

- the initial concentration of phage L_o in solution increases
- the initial concentration of cell receptors R_o available for binding increases
- the normalized forward reaction rate $\alpha = lk_f/k_{int}$ increases (i.e., the valency l of the phage increases and the forward reaction rate k_f increases for a given k_{int})
- the normalized reverse reaction rate $\beta = k_r/k_{int}$ decreases (i.e., the reverse reaction rate k_r decreases for a given k_{int}).

Also, it is interesting to analyze the ratio between the internalized $E(t)$ and bound $C(t)$ phages, which derives from Eq. (6.9) as:

$$\frac{E(\tau)}{C(\tau)} = \frac{1}{\beta + 1} \left(\frac{(\beta + 1)\tau}{1 - e^{-(\beta+1)\tau}} - 1 \right) \tag{6.10}$$

This ratio is interestingly independent of the parameter $\alpha = lk_f/k_{int}$ and steadily grows with time and β.

As an example, the variation of $C(t)$ and $E(t)$ is presented in Fig. 6.3 for $\alpha = 1$ and $\beta = 0.1$, 1, and 10 corresponding to $k_r < k_{int}$, $k_r = k_{int}$, and $k_r > k_{int}$. The normalized concentration of bound and internalized phages grows over time and

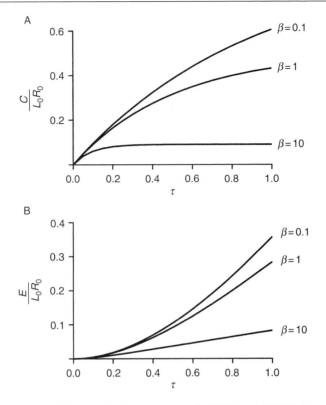

Figure 6.3. Variation of the normalized concentration for (A) bound ($C/(L_oR_o)$) and (B) internalized ($E/(L_oR_o)$) phages as a function of the dimensionless time τ, for $\alpha=1$ and $\beta=0.1$, 1, and 10.

is significantly affected by the parameter β. In particular, a decrease in β (i.e., increase in internalization rate over the reverse rate for binding) is associated with an increase in both complex formation (C) and internalized complexes (E).

Another important deduction from Eq. (6.9) follows when $k_{int}=0$, for which phage only binds receptors on the cell membrane without being internalized. In this case, from Eq. (6.3), it can be readily derived that

$$\frac{C(t)}{L_oR_o} = \frac{1}{K_d}[1-e^{-k_r t}] \quad \text{and} \quad \frac{C_\infty}{L_oR_o} = \frac{1}{K_d} \quad \text{with} \quad K_d = \frac{k_r}{lk_f} \tag{6.11}$$

where K_d is the dissociation constant for the phage–receptor bond, and C_∞ is the concentration of bound phage at equilibrium. Eq. (6.11) demonstrates that the ratio between the concentration of bound (C) and initially injected (L_o) phage is proportional to the number of receptors available (R_o) and inversely proportional to K_d.

Clearly, the conclusions drawn from Eqs. (6.9)–(6.11) hold in the limits of Eq. (6.2), that is to say for sufficiently large initials concentrations of phages and receptors.

During *in vivo* biopanning, phage concentration will be very low in comparison to receptor concentration. As a numerical example, when considering a phage solution with a phage titer of 10^{10} TU/μl, this translates to a concentration of 1.6×10^{-8} M or 16 nM. In the context of *in vivo* screenings, the phage solution will instantaneously be diluted after systemic administration by approximately two orders of magnitude, rendering the phage concentration in the subnanomolar range. If one then accounts for a diversity of $\sim 10^6$–10^9 different phage clones, specific phage concentrations are on the order of fM to aM. *In vivo*, with three-dimensional structured tissue and thus increased cell surface exposure, the number of accessible receptors is maximized. In the case of selecting phage with high affinities, both R_o and $1/K_d$ would be large, leading to significant C/L_o values despite the small initial concentration of free phage in solution. We could define the ratio of the number of complexes formed and initial phage concentration (C/L_o) as the efficiency of a phage. Clearly, this value would be much larger during an *in vivo* biopanning compared to an *in vitro* selection procedure because of the larger R_o (regardless of K_d).

B. Multivalency

Multivalent ligand–receptor interactions are defined as the simultaneous binding of multiple ligands on the same molecule to multiple receptors (Mammen *et al.*, 1998). Biological systems make extensive use of multivalent interactions, which have characteristics that monovalent interactions do not. An example of multivalent ligand–receptor interactions in nature is the attachment of viruses to host cells, involving the simultaneous binding of multiple ligands on the surface of the virus to several receptors on the surface of host cells (Mammen *et al.*, 1998). In the case of phage display, the ligand is fused to the amino terminus of the pIII minor coat protein, which allows for a maximum valency of up to five peptides per phage (Barbas, 2001). Apparent K_d values of synthesized single peptides derived from biopannings often show weaker binding to target receptors than their respective higher affinity phage (Caravan *et al.*, 2007; Helms *et al.*, 2009). This suggests that the selection of peptide ligands using phage libraries is not only relying on the specific targeting moiety displayed on the phage but is also dependent on the presentation of multiple copies, capable of interacting with the target receptor in a multivalent fashion. Indeed, in a recent study a peptide identified from phage display was fused onto a phage mimicking pentavalent dendrimer (Helms *et al.*, 2009). Using this system, it was demonstrated that the binding affinity of a single peptide to its target receptor could be increased by two orders of magnitude using multivalent display (Helms *et al.*, 2009). The major advantage of multivalent interactions is that they can be much stronger than their corresponding monovalent interactions (Mammen *et al.*, 1998). The magnitude of the total free energy resulting from multivalent ligand–receptor binding is high which results in favorable binding affinities (Mammen *et al.*, 1998). In addition, it is hypothesized that multivalent ligands can interact with cell surface receptors via several possible mechanisms (Gestwicki *et al.*, 2002). These include conformational stabilization, statistical rebinding, and clustering of the receptors on the cell membrane (Gestwicki *et al.*, 2002). Conversely, monovalent ligands may only have access to a limited number of binding mechanisms. These ligands typically bind to a single receptor or, less commonly, cause receptor dimerization (Gestwicki *et al.*, 2002).

Stochastic modeling studies comparing multi- with monovalent phages display systems *in vitro* using immobilized recombinant cell surface receptors led to the following observations (Levitan, 1998). Multivalent phage can survive during selection more easily compared to monovalent phage when using the same target concentration, supporting the notion of increased binding affinity. In addition, multivalent phage binding requires that the distance between the target receptor molecules is sufficiently small. Therefore, it can be assumed that multivalent phage binding *in vitro* may depend primarily on the distance/proximity, rather than the number of targets. In contrast, monovalent phage binding is directly correlated to the number of target molecules.

The concept of multivalency is commonly encountered in biological systems. It can be hypothesized that multivalency significantly contributes to the successful use of phage display *in vivo*. Cell surface receptors, the binding partners of phage, play a major role in signal transduction (Kiessling *et al.*, 2006). It can be conceived that many of these receptors do not operate as isolated single entities, but rather as part of oligomeric macromolecular complexes. Phage, as a multivalent ligand, has the necessary attributes to recognize multiprotein complexes, to induce receptor clustering, and/or to alter the proximity between the receptors (Kiessling *et al.*, 2006). These explanations certainly remain speculative but given the extraordinary targeting ability of phage observed in our *in vivo* biopannings, one can assume that the multivalent architecture of the phage particle may be a key contributor in the successful selection of novel ligand–receptor pairs *in vivo*.

C. Biology driven effects within tissue microenvironments

For receptor-targeted strategies to be successful, it is critical that the targeted receptor is expressed in high enough numbers. This was elegantly demonstrated in a study by Park *et al.* (2002) that evaluated the antitumor efficacy of HER2-targeted immunoliposomes loaded with Doxorubicin in breast cancer xenograft models expressing different levels of HER2-receptors (range from $\sim 10^3$ to $\sim 10^6$ receptors/cell). Antitumoral efficacy was found in all tested animal models, except for the mice bearing the xenografts with only $\sim 10^3$ receptors/cell;

indicating there is a threshold for mediating efficient binding of the targeted liposomes to the receptor (Park *et al.*, 2002). This finding further explains selectivity for tissues overexpressing receptors versus tissues that only express moderate levels of the same receptor. However, *in vivo* phage display maps the tissue microenvironment, with the target receptor present in its relevant microanatomical context. This adds a further level of specificity that goes beyond receptor-expression levels, in that *in vivo* phage display maps out receptors uniquely accessible from the circulation (Ozawa *et al.*, 2008). The microenvironment will have an influence on the expression of receptors that might not be present on similar cell-types in other tissues and it might rearrange proteins from luminal or abluminal regions of the cell to the cell-surface. Lastly, pathological tissues have cell-types associated in their microenvironment that would normally not occur, such as tumor-associated monocytes (Ozawa *et al.*, 2008; Qian and Pollard, 2010).

III. SYNERGISTIC EFFECTS INTRINSIC TO PHAGE

In vivo phage display screening likely benefits from interactions mediated by phage properties, that is phage-specific effects that are not peptide-insert specific. The filamentous bacteriophage used in our laboratory for *in vivo* phage display (fd), has a length of $L_o \sim 900$ nm and a diameter of ~ 7 nm (6.6 nm) (Wang *et al.*, 2006). In this case, phage will be considered to be a particle with nanodimensions and the phage-specific, or intrinsic effects and interactions are aiding in the increase of the effective (or local) peptide concentration (which would increase reaction rate), increase the probability of interaction, and/or minimize the free energy of this step, so that specific peptide-receptor recognition and internalization can take place. Such interactions intrinsic to phage include the following sections.

A. Intermolecular forces, such as electrostatic forces, van der Waals forces, and hydrophobic forces—these forces can be described in terms of adsorption isotherms

Of the intermolecular forces, it has been published (Goncalves *et al.*, 2005; Ziegler *et al.*, 2003) that the electrostatic interaction within biological systems (peptide–membrane interactions) can contribute ~ 30–70% to the total free energy of a biological process. The ~ 2700 copies of the major coat protein pVIII constitute about 99% of the entire protein mass of the phage-particle (Wiseman *et al.*, 1976). This coat protein consists of 50 amino acids, which form an α-helix with a large hydrophobic region between the protein-termini (Nakashima *et al.*, 1975). The hydrophobic regions interact with the neighboring pVIII proteins forming a protein sheath conferring resistance of phage to harsh conditions such as high salt concentration, acidic pH, chaotropic agents, and prolonged storage. Titration analysis of the ionic properties of the phage particle revealed a total number of 8800 charged residues at pH 7.4, leading to a surface charge density of $0.46 \, eq/nm^2$, making it very feasible that electrostatic interactions are a major contributor to the intrinsic phage-effects (Zimmermann *et al.*, 1986).

B. Diffusion and extravasation, especially through fenestrated vasculature

Targeting of organs or diseased vasculature can be generally achieved through two strategies: active and passive targeting. Active targeting through phage is mediated by a peptide displayed on the pIII protein and is therefore based on a specific ligand–receptor interaction. In passive targeting the intrinsic properties of the phage particle itself could contribute to increased target/nontarget ratio at a site of interest. It is not without precedent that polymers and nanoparticles have access to subcellular tissue compartments. Polymers such as poly(ethylene glycol) (PEG) seem to be required for efficiency in many existing delivery systems and, ultimately, the phage-particle may well behave as such a particle (Geng *et al.*, 2007). The landmark of passive targeting is a phenomenon named enhanced permeability and retention (EPR), a tumor targeting strategy initially proposed by Matsumura and Maeda (1986). The EPR phenomenon is based on the observation that tumors possess unique characteristics that are absent in normal tissues, such as extensive angiogenesis, defective vascular beds, impaired lymphatic drainage/recovery system, and highly increased production of permeability factors (Matsumura and Maeda, 1986). This causes the tumor-associated neovasculature to be highly permeable, allowing the leakage of circulating macromolecules and nanoparticles into the tumor interstitium. This effect is widely being exploited in anticancer strategies like gene and drug delivery, molecular imaging and antibody therapy, using drug–polymer conjugates, micelles, liposomes, nanoparticles, DNA polyplexes, and lipid particles (Fang *et al.*, 2003; Iyer *et al.*, 2006; Maeda, 2001a, b; Maeda *et al.*, 2009). Recently, studies using *Lactobacillus* sp. and *Salmonella typhimurium* suggested that the EPR effect functions also for bacteria larger than 1000 nm (Hoffman, 2009; Zhao et al., 2005, 2006). A combination of tumor cell-specific targeting by receptor-mediated internalization and EPR accumulation (i.e., the integration of active and passive targeting) may enable targeting of receptors even if they are expressed at low concentration at the site of interest and it may be envisioned that ligand-targeted phage could be exploiting EPR to enhance its targeting capabilities. Furthermore, this phenomenon may suggest the importance of conjugating chemically synthesized peptides to polymers or nanoparticles rather than using the peptide by itself in order to evoke some of these properties.

C. Effects on the pharmacokinetics of the targeted system

Peptides in solution possess biological activities that render them powerful therapeutic and targeting agents. However, their use *in vivo* is often hampered by low serum stability and fast elimination from the circulation. In contrast, the display of peptides on phage may lead to increased longevity in the circulation, thus increasing the chance of interaction and allowing for better accumulation in tissues of interest. This becomes of even higher importance when the vascular volume of the targeted tissue is relatively small compared to the complete host vasculature, that is, for poorly perfused organs. Park *et al.* (2009) performed a systematic evaluation of tumor-targeting with nanoparticles differing in ligand density, target-receptor, and shape (elongated vs. spherical). They concluded that it is critical to optimize the pharmacokinetic behavior of a targeted system in order to maximize accumulation in target-tissue.

It has been shown that serum half-life of filamentous phage can vary greatly based on animal-species and model (Zou *et al.*, 2004), and comparisons between literatures are hampered by different phage-detection methods. However, Molenaar *et al.* (2002) performed pharmacokinetic studies with metabolically radiolabeled ^{35}S-methionine/cysteine wild-type M13 phage so that both uninfectious and infectious phages could readily be detected. The study reports a serum half-life of 4.5 h for untargeted filamentous phage. The reported half-life dropped 1–2 orders of magnitude after the major phage coat was chemically modified with targeting moieties to stimulate uptake by galactose recognizing hepatic receptors or scavenger receptors. However, this experimental condition is vastly different from the typical scenario, where a smaller number of targeting moieties are displayed recombinantly on pIII. It is not inconceivable that chemical modification enhances nonspecific uptake of phage particles and indeed this has been reported for ^{125}I labeled phage (Johns *et al.*, 2000).

D. Effects of size, shape, and rigidity

An increasing amount of research in the field of nanomedicine is focused on optimizing the size, shape, and surface properties of particles for downstream applications (Decuzzi and Ferrari, 2008; Decuzzi *et al.*, 2005, 2006, 2009; Tasciotti *et al.*, 2008). The extraordinary ability of peptides displayed on filamentous phage to sense biological diversities among vascular beds with high specificity depends also on their size, elongated shape, and mechanical properties. Indeed, polymer micelle assemblies (filomicelles) mimicking the size and shape of natural filamentous viruses were successfully used as drug-delivery vehicles for cancer therapy in tumor-bearing mice. It was demonstrated that mice treated with either free drug or drug-loaded filomicelles showed a clear length depended inhibition of tumor growth when using drug-loaded filomicelles compared to the free drug (Geng et al., 2007).

The phage particle has a diameter which is comparable with the characteristic size of receptor molecules expressed along the endothelial cell wall. The particle exhibits an elongated cylindrical shape with a high slenderness ratio $\varepsilon \sim 4 \times 10^{-3}$ ($\varepsilon = d/2L_o$). Some of the mechanical properties of the M13 phage capsid have been recently characterized using an optical trap with nanometer-scale position resolution (Khalil et al., 2007), which has allowed to more accurately estimate the persistence length l_p of the filament and its elastic stretching modulus y per unit length. In particular, a single phage particle was stretched to derive its force–displacement curve, and a resultant persistence length $l_p = 1265.7 \pm 220.4$ nm was found.

The persistence length of the phage particle is a basic mechanical property that quantifies the stiffness or rigidity of the polymeric capsid and could represent another phage-inherent feature enabling successful in vivo biopanning and vascular mapping. For phage with length L_o smaller than the persistence length l_p ($= 1265.7 \pm 220.4$ nm), the phage can be considered as a flexible rod with a modulus $Y \sim 50$ MPa ($= 16l_p(k_BT)/(\pi d^2)$). Interestingly, filamentous bacteriophage is more rigid than single DNA strands ($l_p = 50$ nm $\ll L_{DNA}$ (Woolley and Kelly, 2001)) and "softer" than microtubules ($l_p = 1$ mm $\gg L_o$ for microtubulus (Li et al., 2002)).

The dynamics of rods in laminar flows is strongly affected by their flexibility. Tornberg and Shelley (2004) introduced an effective viscosity parameter Π to measure the ratio between the hydrodynamic forces exerted over the flexible rod and the elastic forces associated with rod deformation under flow. The effective viscosity parameter Π is defined as:

$$\Pi = \frac{8\pi\mu SL_o^2}{\left(Y \frac{\pi d^4}{64}\right)\frac{1}{L_o^2}} \tag{6.12}$$

where S is the wall shear rate and μ is the fluid viscosity ($\mu = 10^{-3}$ Pa s for water solution). Small values of Π imply that the filament is stiff enough not to be deformed under flow and can therefore be considered as rigid. For the phage ($L_o \sim 900$ nm; $Y \sim 50$ MPa) and considering a shear rate at the wall as large as $S = 10^3$ s^{-1}, the effective viscosity Π would be of the order of 10^3. Smaller values of Π are expected in the tumor vasculature, where the wall shear rate S is generally smaller than 10^3 s^{-1}. This would imply that the hydrodynamic forces could slightly affect the configuration of the phage particle in a physiological laminar flow; whereas in the tumor vasculature the phage behavior would tend to that of a rigid rod.

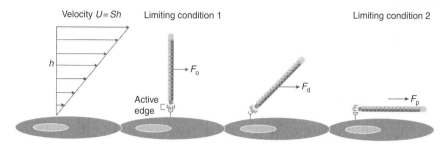

Figure 6.4. Phage modeled as a rigid rod moving in the "pole vaulting" motion in a laminar capillary flow exerting a force F_o when positioned orthogonal to the direction of the flow and F_p when positioned parallel to the direction of the flow. By moving in this way, active edge accessibility to cell surface receptors is maximized. The velocity profile of the flow is shown as a function of h (distance from the cell membrane) and S (wall sheer rate).

Stover and Cohen (1990) observed that rigid rods moving in the vicinity of a wall exhibit a "pole vaulting" motion during which the rod edge in close proximity to the wall acts as the pivot around which the rod rotates at the same time as it translates following the flow (Fig. 6.4). This process is repeated periodically along the direction of flow. For a phage (flexible rod), the "pole vaulting" motion could be accompanied also by some degree of longitudinal compression/tension and bending, as described experimentally by Forgacs and Mason (1959a, b). Clearly the "pole vaulting" motion would increase the likelihood of interaction of the "active edge" of the bacteriophage (in most cases pIII) with the endothelial cells.

Within tumor vasculature, the phage could be considered as a rigid rod subjected to a dislodging hydrodynamic force exerted in the direction of the flow depending slightly on the phage orientation. The dislodging hydrodynamic forces exerted over a rigid rod positioned orthogonally (F_o—limiting condition 1 in Fig. 6.4) and parallelly (F_p—limiting condition 2 in Fig. 6.4) to the vascular wall take the form (De Mestre and Russel, 1975):

$$\begin{aligned} F_p &= 2\pi\mu U\omega L_o[2 + \omega(W_p - 0.614)]; \\ F_o &= 4\pi\mu U\omega l[2 + \omega(W_o - 1.386)]; \end{aligned} \tag{6.13}$$

with

$$\begin{aligned} W_p &= 2\sin^{-1}h[\chi] - 3(1 + \chi^{-2})^{1/2} + \frac{7}{2\chi} - \frac{1}{2\chi^2(1 + \chi^{-2})^{1/2}} \\ W_o &= (1 + \chi^{-1})\log(1 + \chi) + (\chi^{-1} - 1)\log(1 - \chi) + \frac{\chi}{2} \end{aligned} \tag{6.14}$$

where $\omega = \log[2/\varepsilon]^{-1}$, $U = Sh$, and $\chi = L_o/h$, h being the separation distance between the phage center and the wall. Using Eqs. (6.13) and (6.14), for a phage particle in close proximity to the vessel wall ($h = 10$ nm) and $S = 10^3$ s^{-1}; the dislodging forces result in $F_p \sim 40$ fN and $F_o \sim 50$ fN. The implication of this numerical example is two-fold: (i) the drag force for a given fixed separation distance is relatively insensitive to the orientation of the phage with F_p being only slightly smaller than F_o ($\sim 20\%$); (ii) the hydrodynamic dislodging forces are at least one order of magnitude smaller than the rupture forces measured for single ligand–receptor bonds (Evans and Calderwood, 2007; Merkel et al., 1999). Bond rupture forces have been characterized for many different molecules using different force spectrometry approaches (Evans and Calderwood, 2007; Merkel et al., 1999). The strength of such bonds not only depends on the ligand–receptor family but also is strongly affected by the loading rate or, in other words, by how fast the load is transferred to the ligand–receptor bond. Considering a drag force exerted over the phage ranging between 10 and 100 fN (see above), a wall shear rate in the order of 10^3–10^4 s^{-1}, the loading rate associated with the phage would range between 10 and 10^3 pN/s. Under these conditions, the classical endothelial receptor ICAM-1 would unbind from the corresponding ligand LFA-1 for a force of the order of 10 pN (Zhang et al., 2002). Similar or even larger rupture forces would be expected for a selected phage-peptide.

Evidently, this situation is in favor of phage remaining bound to a cognate receptor and it is important to note that differences between F_p and F_o would be larger for particles with nonrod-like shapes (e.g., a prolate spheroid or ellipsoids) and particles with a smaller slender ratio (e.g., shorter phage). The small dependence of the drag force on the phage orientation mediated by the unique phage-shape and the slender ratio favors the binding with the potential wall receptors regardless of the phage orientation as long as a ligand molecule on the phage's "active edge" is exposed to the endothelium.

In addition to the drag force, phage particles that are attaching with their "active edge" to the endothelial wall under flow would also feel a couple that would tend to rotate and push them down on the cell membrane. Given the flexibility of the "active edge" of the phage and the small dislodging forces compared with the adhesive interactions, the bacteriophage would rotate and lay down on the cell membrane without detaching from the vascular target (Fig. 6.2). Under these conditions, the phage would be quickly enveloped and internalized by the cell membrane, which indeed would be exposed to a long cylindrical particle with a circular cross-section with a small diameter d. Indeed, Decuzzi and Ferrari (2008) have shown that cylinders with circular cross-sections are more easily wrapped by the cell membrane and eventually more quickly internalized.

IV. CONCLUSION

Identification of druggable targets through *in vivo* combinatorial phage library selection has already resulted in several successful new therapeutic agents (Arap *et al.*, 2002; Cardo-Vila *et al.*, 2010; Giordano *et al.*, 2010; Hajitou *et al.*, 2006). We have discussed a number of aspects of ligand–receptor mediated targeting and intrinsic phage characteristics that likely contribute to the success of *in vivo* phage screenings. By further study and characterization of these topics, there is the potential to enhance phage physical properties for optimal delivery and targeting. These studies may ultimately lead to an overall improvement of phage display derived therapeutic strategies.

References

Arap, W., Kolonin, M. G., Trepel, M., Lahdenranta, J., Cardo-Vila, M., Giordano, R. J., Mintz, P. J., Ardelt, P. U., Yao, V. J., Vidal, C. I., Chen, L., Flamm, A., *et al.* (2002). Steps toward mapping the human vasculature by phage display. *Nat. Med.* **8**(2), 121–127.

Barbas, C., III (2001). *In* "Phage Display: A Laboratory Manual" (C. Barbas III, ed.). Cold Spring Harbor Laboratory Press, Cold Spring Harbor, New York.

Brown, D. M., Pellecchia, M., and Ruoslahti, E. (2004). Drug identification through in vivo screening of chemical libraries. *Chembiochem* **5**(6), 871–875.

Caravan, P., Das, B., Dumas, S., Epstein, F. H., Helm, P. A., Jacques, V., Koerner, S., Kolodziej, A., Shen, L., Sun, W. C., and Zhang, Z. (2007). Collagen-targeted MRI contrast agent for molecular imaging of fibrosis. *Angew. Chem. Int. Ed. Engl.* **46**(43), 8171–8173.

Cardo-Vila, M., Giordano, R. J., Sidman, R. L., Bronk, L. F., Fan, Z., Mendelsohn, J., Arap, W., and Pasqualini, R. (2010). From combinatorial peptide selection to drug prototype (II): Targeting the epidermal growth factor receptor pathway. *Proc Natl Acad Sci USA* **107**(11), 5118–5123.

Christianson, D. R., Ozawa, M. G., Pasqualini, R., and Arap, W. (2007). Techniques to decipher molecular diversity by phage display. *Methods Mol. Biol.* **357**, 385–406.

De Mestre, N. J., and Russel, W. B. (1975). Low-Reynolds-number translation of a slender cylinder near a plane wall. *J. Eng. Math.* **9**(2), 81–91.

Decuzzi, P., and Ferrari, M. (2008). The receptor-mediated endocytosis of nonspherical particles. *Biophys. J.* **94**(10), 3790–3797.

Decuzzi, P., Lee, S., Bhushan, B., and Ferrari, M. (2005). A theoretical model for the margination of particles within blood vessels. *Ann. Biomed. Eng.* **33**(2), 179–190.

Decuzzi, P., Causa, F., Ferrari, M., and Netti, P. A. (2006). The effective dispersion of nanovectors within the tumor microvasculature. *Ann. Biomed. Eng.* **34**(4), 633–641.

Decuzzi, P., Pasqualini, R., Arap, W., and Ferrari, M. (2009). Intravascular delivery of particulate systems: Does geometry really matter? *Pharm. Res.* **26**(1), 235–243.

Dickson, M., and Gagnon, J. P. (2004). Key factors in the rising cost of new drug discovery and development. *Nat. Rev. Drug. Discov.* **3**(5), 417–429.

Evans, E. A., and Calderwood, D. A. (2007). Forces and bond dynamics in cell adhesion. *Science* **316**(5828), 1148–1153.

Fang, J., Sawa, T., and Maeda, H. (2003). Factors and mechanism of "EPR" effect and the enhanced antitumor effects of macromolecular drugs including SMANCS. *Adv. Exp. Med. Biol.* **519**, 29–49.

Forgacs, O. L., and Mason, S. G. (1959a). Particle motions in sheared suspensions: IX. Spin and deformation of threadlike particles. *J. Colloid Sci.* **14**(5), 457–472.

Forgacs, O. L., and Mason, S. G. (1959b). Particle motions in sheared suspensions: X. Orbits of flexible threadlike particles. *J. Colloid Sci.* **14**(5), 473–491.

Geng, Y., Dalhaimer, P., Cai, S., Tsai, R., Tewari, M., Minko, T., and Discher, D. E. (2007). Shape effects of filaments versus spherical particles in flow and drug delivery. *Nat. Nanotechnol.* **2**(4), 249–255.

Gestwicki, J. E., Cairo, C. W., Strong, L. E., Oetjen, K. A., and Kiessling, L. L. (2002). Influencing receptor-ligand binding mechanisms with multivalent ligand architecture. *J. Am. Chem. Soc.* **124**(50), 14922–14933.

Giordano, R. J., Cardo-Vila, M., Salameh, A., Anobom, C. D., Zeitlin, B. D., Hawke, D. H., Valente, A. P., Almeida, F. C., Nor, J. E., Sidman, R. L., Pasqualini, R., and Arap, W. (2010). From combinatorial peptide selection to drug prototype (I): Targeting the vascular endothelial growth factor receptor pathway. *Proc. Natl. Acad. Sci. USA* **107**(11), 5112–5117.

Goncalves, E., Kitas, E., and Seelig, J. (2005). Binding of oligoarginine to membrane lipids and heparan sulfate: Structural and thermodynamic characterization of a cell-penetrating peptide. *Biochemistry* **44**(7), 2692–2702.

Hajitou, A., Pasqualini, R., and Arap, W. (2006). Vascular targeting: Recent advances and therapeutic perspectives. *Trends Cardiovasc. Med.* **16**(3), 80–88.

Helms, B. A., Reulen, S. W., Nijhuis, S., de Graaf-Heuvelmans, P. T., Merkx, M., and Meijer, E. W. (2009). High-affinity peptide-based collagen targeting using synthetic phage mimics: From phage display to dendrimer display. *J. Am. Chem. Soc.* **131**(33), 11683–11685.

Hoffman, R. M. (2009). Tumor-targeting amino acid auxotrophic *Salmonella typhimurium*. *Amino Acids* **37**(3), 509–521.

Houghten, R. A., Pinilla, C., Giulianotti, M. A., Appel, J. R., Dooley, C. T., Nefzi, A., Ostresh, J. M., Yu, Y., Maggiora, G. M., Medina-Franco, J. L., Brunner, D., and Schneider, J. (2008). Strategies for the use of mixture-based synthetic combinatorial libraries: Scaffold ranking, direct testing in vivo, and enhanced deconvolution by computational methods. *J. Comb. Chem.* **10**(1), 3–19.

Iyer, A. K., Khaled, G., Fang, J., and Maeda, H. (2006). Exploiting the enhanced permeability and retention effect for tumor targeting. *Drug Discov. Today* **11**(17–18), 812–818.

Johns, M., George, A. J., and Ritter, M. A. (2000). In vivo selection of sFv from phage display libraries. *J. Immunol. Methods* **239**(1–2), 137–151.

Khalil, A. S., Ferrer, J. M., Brau, R. R., Kottmann, S. T., Noren, C. J., Lang, M. J., and Belcher, A. M. (2007). Single M13 bacteriophage tethering and stretching. *Proc. Natl. Acad. Sci. USA* **104**(12), 4892–4897.

Kiessling, L. L., Gestwicki, J. E., and Strong, L. E. (2006). Synthetic multivalent ligands as probes of signal transduction. *Angew. Chem. Int. Ed. Engl.* **45**(15), 2348–2368.

Kolonin, M. G., Sun, J., Do, K. A., Vidal, C. I., Ji, Y., Baggerly, K. A., Pasqualini, R., and Arap, W. (2006). Synchronous selection of homing peptides for multiple tissues by in vivo phage display. *FASEB J.* **20**(7), 979–981.

Krag, D. N., Shukla, G. S., Shen, G. P., Pero, S., Ashikaga, T., Fuller, S., Weaver, D. L., Burdette-Radoux, S., and Thomas, C. (2006). Selection of tumor-binding ligands in cancer patients with phage display libraries. *Cancer Res.* **66**(15), 7724–7733.

Lesko, L. J. (2007). Paving the critical path: How can clinical pharmacology help achieve the vision? *Clin. Pharmacol. Ther.* **81**(2), 170–177.

Levitan, B. (1998). Stochastic modeling and optimization of phage display. *J. Mol. Biol.* **277**(4), 893–916.

Li, H., DeRosier, D. J., Nicholson, W. V., Nogales, E., and Downing, K. H. (2002). Microtubule structure at 8 A resolution. *Structure* **10**(10), 1317–1328.

Maeda, H. (2001a). The enhanced permeability and retention (EPR) effect in tumor vasculature: The key role of tumor-selective macromolecular drug targeting. *Adv. Enzyme Regul.* **41**, 189–207.

Maeda, H. (2001b). SMANCS and polymer-conjugated macromolecular drugs: Advantages in cancer chemotherapy. *Adv. Drug Deliv. Rev.* **46**(1–3), 169–185.

Maeda, H., Bharate, G. Y., and Daruwalla, J. (2009). Polymeric drugs for efficient tumor-targeted drug delivery based on EPR-effect. *Eur. J. Pharm. Biopharm.* **71**(3), 409–419.

Mammen, M., Choi, S.-K., and Whitesides, G. M. (1998). Polyvalent interactions in biological systems: Implications for design and use of multivalent ligands and inhibitors. *Angew. Chem. Int. Ed. Engl.* **37**(20), 2754–2794.

Matsumura, Y., and Maeda, H. (1986). A new concept for macromolecular therapeutics in cancer chemotherapy: Mechanism of tumoritropic accumulation of proteins and the antitumor agent smancs. *Cancer Res.* **46**(12 Pt 1), 6387–6392.

Merkel, R., Nassoy, P., Leung, A., Ritchie, K., and Evans, E. (1999). Energy landscapes of receptor-ligand bonds explored with dynamic force spectroscopy. *Nature* **397**(6714), 50–53.

Molenaar, T. J., Michon, I., de Haas, S. A., van Berkel, T. J., Kuiper, J., and Biessen, E. A. (2002). Uptake and processing of modified bacteriophage M13 in mice: Implications for phage display. *Virology* **293**(1), 182–191.

Nakashima, Y., Wiseman, R. L., Konigsberg, W., and Marvin, D. A. (1975). Primary structure and sidechain interactions of PFL filamentous bacterial virus coat protein. *Nature* **253**(5486), 68–71.

Ozawa, M. G., Zurita, A. J., Dias-Neto, E., Nunes, D. N., Sidman, R. L., Gelovani, J. G., Arap, W., and Pasqualini, R. (2008). Beyond receptor expression levels: The relevance of target accessibility in ligand-directed pharmacodelivery systems. *Trends Cardiovasc. Med.* **18**(4), 126–132.

Park, J. W., Hong, K., Kirpotin, D. B., Colbern, G., Shalaby, R., Baselga, J., Shao, Y., Nielsen, U. B., Marks, J. D., Moore, D., Papahadjopoulos, D., and Benz, C. C. (2002). Anti-HER2 immunolipo-somes: Enhanced efficacy attributable to targeted delivery. *Clin. Cancer Res.* **8**(4), 1172–1181.

Park, J. H., von Maltzahn, G., Zhang, L., Derfus, A. M., Simberg, D., Harris, T. J., Ruoslahti, E., Bhatia, S. N., and Sailor, M. J. (2009). Systematic surface engineering of magnetic nanoworms for in vivo tumor targeting. *Small* **5**(6), 694–700.

Pasqualini, R., and Ruoslahti, E. (1996). Organ targeting in vivo using phage display peptide libraries. *Nature* **380**(6572), 364–366.

Pasqualini, R., Moeller, B. J., and Arap, W. (2010). Leveraging molecular heterogeneity of the vascular endothelium for targeted drug delivery and imaging. *Semin. Thromb. Hemost.* **36**(3), 343–351.

Qian, B. Z., and Pollard, J. W. (2010). Macrophage diversity enhances tumor progression and metastasis. *Cell* **141**(1), 39–51.

Smith, G. P., and Petrenko, V. A. (1997). Phage display. *Chem. Rev.* **97**(2), 391–410.

Stover, C. A., and Cohen, C. (1990). The motion of rodlike particles in the pressure-driven flow between two flat plates. *Rheol. Acta* **29**, 192.

Tasciotti, E., Liu, X., Bhavane, R., Plant, K., Leonard, A. D., Price, B. K., Cheng, M. M., Decuzzi, P., Tour, J. M., Robertson, F., and Ferrari, M. (2008). Mesoporous silicon particles as a multistage delivery system for imaging and therapeutic applications. *Nat. Nanotechnol.* **3**(3), 151–157.

Tornberg, and Shelley (2004). Simulating the dynamics and interactions of flexible fibers in stokes flows. *J. Comput. Phys.* **196**, 8–40.

Trepel, M., Pasqualini, R., and Arap, W. (2008). Chapter 4: Screening phage-display peptide libraries for vascular targeted peptides. *Methods Enzymol.* **445**, 83–106.

Wang, Y. A., Yu, X., Overman, S., Tsuboi, M., Thomas, G. J., Jr., and Egelman, E. H. (2006). The structure of a filamentous bacteriophage. *J. Mol. Biol.* **361**(2), 209–215.

Wiseman, R. L., Berkowitz, S. A., and Day, L. A. (1976). Different arrangements of protein subunits and single-stranded circular DNA in the filamentous bacterial viruses fd and Pf1. *J. Mol. Biol.* **102**(3), 549–561.

Woodcock, J., and Woosley, R. (2008). The FDA critical path initiative and its influence on new drug development. *Annu. Rev. Med.* **59**, 1–12.

Woolley, A. T., and Kelly, R. T. (2001). Deposition and characterization of extended SIngle-stranded DNA molecules on surfaces. *Nano Lett.* **1**(7), 345–348.

Zhang, X., Wojcikiewicz, E., and Moy, V. T. (2002). Force spectroscopy of the leukocyte function-associated antigen-1/intercellular adhesion molecule-1 interaction. *Biophys. J.* **83**(4), 2270–2279.

Zhao, M., Yang, M., Li, X. M., Jiang, P., Baranov, E., Li, S., Xu, M., Penman, S., and Hoffman, R. M. (2005). Tumor-targeting bacterial therapy with amino acid auxotrophs of GFP-expressing *Salmonella typhimurium*. *Proc. Natl. Acad. Sci. USA* **102**(3), 755–760.

Zhao, M., Yang, M., Ma, H., Li, X., Tan, X., Li, S., Yang, Z., and Hoffman, R. M. (2006). Targeted therapy with a Salmonella typhimurium leucine-arginine auxotroph cures orthotopic human breast tumors in nude mice. *Cancer Res.* **66**(15), 7647–7652.

Ziegler, A., Blatter, X. L., Seelig, A., and Seelig, J. (2003). Protein transduction domains of HIV-1 and SIV TAT interact with charged lipid vesicles. Binding mechanism and thermodynamic analysis. *Biochemistry* **42**(30), 9185–9194.

Zimmermann, K., Hagedorn, H., Heuck, C. C., Hinrichsen, M., and Ludwig, H. (1986). The ionic properties of the filamentous bacteriophages Pf1 and fd. *J. Biol. Chem.* **261**(4), 1653–1655.

Zou, J., Dickerson, M. T., Owen, N. K., Landon, L. A., and Deutscher, S. L. (2004). Biodistribution of filamentous phage peptide libraries in mice. *Mol. Biol. Rep.* **31**(2), 121–129.

7

Strategies for Targeting Tumors and Tumor Vasculature for Cancer Therapy

Prashanth Sreeramoju and Steven K. Libutti

Montefiore Medical Center, Albert Einstein College of Medicine, Bronx, New York, USA

ABSTRACT

Effective cancer therapy remains a challenge despite recent advances in the identification of novel targets. A major limitation of most chemotherapeutic drugs is their systemic toxicity and the efficacy of cancer treatments is, by and large, determined by the ability to balance their benefits against their toxicity. Targeted treatments for cancer, especially those that target the tumor vasculature, have provided promising antitumor results with minimal systemic toxicity. To date significant progress has been made in developing a variety of delivery systems to target cancer and its vasculature ranging from isolated limb and organ perfusion to tumor targeted biological and nonbiological vectors. © 2010, Elsevier Inc.

Advances in Genetics, Vol. 69
Copyright 2010, Elsevier Inc. All rights reserved.

0065-2660/10 $35.00
DOI: 10.1016/S0065-2660(10)69015-3

I. INTRODUCTION

Cancer remains the second leading cause of mortality in the United States after heart disease (National Center for Health Statistics and Centers for Disease Control and Prevention, 2006). Cancer affects roughly 1.5 million people in the United States each year with roughly 500,000 deaths annually. A variety of therapies have been developed to treat advanced stage disease over the last five decades. Many of these drugs have had limited success for solid tumors related to either dose-limiting systemic toxicity or the development of tumor resistance. The struggle to maximize efficacy and to minimize systemic toxicity has lead to a paradigm shift in the treatment of cancer.

Increasing knowledge about the tumor and its biology has led to an approach that now targets aspects of the tumor microenvironment in addition to the tumor cell itself. This has lead to the development of new methods for drug delivery targeting tumors and the tumor-associated vasculature. These strategies vary from isolated limb and organ perfusion techniques to systemic delivery targeting tumors and the tumor microenvironment using both biologic and nonbiologic vectors.

II. ISOLATED LIMB PERFUSION

Isolated limb perfusion (ILP) techniques were initially pioneered by Creech and Krementz at Tulane University in New Orleans (Creech et al., 1958). The basic principle behind ILP is to physically confine the delivery of chemotherapy and/or biologics to the limb involved with cancer (usually melanoma or sarcoma) using catheters and tourniquets. This results in dose intensive targeting of the tumor and its vasculature by isolating the chemotherapeutic agent for maximum effect and minimum systemic side effects. This approach can achieve regional concentrations of chemotherapeutic agents 15–25 times higher than what can be reached with systemic administration, but without systemic side effects. Hyperthermia is also often used, as tumor cells are very sensitive to high temperatures (Benckhuijsen et al., 1988).

A variety of chemotherapies and biologics have been delivered using the ILP technique, including melphalan, doxorubicin, tumor necrosis factor (TNF), and cisplatin. The combination of chemotherapy and TNF delivered via ILP impacted the treatment for large unresectable soft tissue sarcoma of the extremity, with a 20–30% complete response rate and a 50% partial response rate (Eggermont et al., 1996a, b, 1999; Gutman et al., 1997; Hill et al., 1993). Limb salvage was achieved in 74–87% of patients (Grünhagen et al., 2006).

The toxicity for ILP is often secondary to local toxicity from the effects of the chemotherapeutic agents on the tissues of the limb. Toxicities are often reversible ranging from slight erythema or edema to extensive epidermolysis and compartment syndrome (Wieberdink *et al.*, 1982). Rarely, leak of chemotherapeutic drugs into the systemic circulation can lead to systemic toxicity. Overall, ILP is well tolerated and can be a very effective therapy for the small percentage of cancers confined to an extremity.

III. ISOLATED LIVER PERFUSION

Isolated hepatic perfusion (IHP) is a regional treatment technique that isolates the vasculature to the liver in order to allow the delivery of high-dose chemotherapy, biologic agents, and hyperthermia directly to unresectable cancers confined to the liver. The liver is the sole or dominant site of metastases from a variety of histologies, including colon cancer and ocular melanoma (Libutti *et al.*, 2000). Regional treatment targeting the liver has the advantage of maximizing drug delivery while minimizing systemic toxicity by allowing significant dose escalation as the liver can recover very quickly from regional toxicities.

The IHP technique is performed during the course of a surgical exploration of the abdomen by isolating the vasculature of the liver and by effecting inflow from the common hepatic artery and gastroduodenal artery and outflow through an isolated segment of inferior vena cava. During isolation, infrahepatic and portal blood is shunted centrifugally to the axillary vein externally (see Fig. 7.1). Isolation can be confirmed using a continuous 131I-labeled serum albumin leak monitoring system (Libutti *et al.*, 2000). This technique has progressively evolved over the years into a minimally invasive approach using percutaneously placed catheters. A double balloon catheter is placed under fluoroscopic guidance into the retrohepatic IVC for venous isolation and targeted drug delivery via a catheter in the hepatic artery (Miao *et al.*, 2008; Pingpank *et al.*, 2005).

The agents studied to date using an IHP or PHP (percutaneous hepatic perfusion) technique are melphalan, doxorubicin, oxaliplatin, and TNF (Alexander *et al.*, 2005). Studies have demonstrated partial response rates of 67% in unresectable liver colorectal metastases and an average overall survival of 16.3 months. These results reflect significant antitumor activity in patients with unresectable hepatic metastases who have progressed through standard systemic chemotherapy (Alexander *et al.*, 2005). IHP is associated with limited toxicity manifested as chemical hepatitis secondary to the effects of chemotherapeutic drugs on the normal hepatic parenchyma. It is seen in almost every patient as a transient and reversible increase in hepatic transaminases and the serum bilirubin.

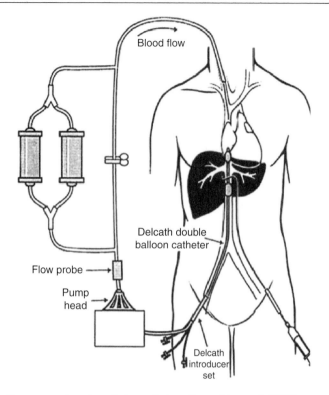

Figure 7.1. Percutaneous hepatic perfusion technique. Source Miao *et al.* (2008).

The major limitation to IHP is the complexity and expense associated with the treatment. The open technique can be performed only once compared to the percutaneous technique, which can be done for as many as eight cycles (Alexander *et al.*, 2005). While response rates to the systemic delivery of newer chemotherapy regimens have been on par with the results from the IHP therapy, in 5-FU (5-flurouracil) refractory patients, these responses are often only transient. The role of both IHP and PHP is still evolving.

IV. CHEMOEMBOLIZATION

Chemoembolization combines the infusion of chemotherapeutic drugs with particle embolization. This results in tumor ischemia and better local drug concentrations and retention. Transarterial chemoembolization (TACE) is performed to control symptoms and as a bridge to transplantation in hepatocellular cancer (HCC) patients. In some indications it can be used to gain local tumor control prior to liver resection. It's been shown that TACE results in local tumor control in 18–70% of cases. The most common single-agents used in TACE are doxorubicin followed by cisplatin and epirubicin. These drugs are complexed with lipiodol (iodized oil)—a vehicle used to carry chemotherapeutic drugs that persists in tumor nodules after injection into selective branches of the hepatic artery (Okayasu *et al.*, 1988). A complication associated with TACE is the postembolization syndrome (PES). This syndrome is characterized by nausea, vomiting, abdominal pain, and fever. PES is seen in 60–70% of patients receiving TACE (Wigmore *et al.*, 2003). Serious complications include hepatic insufficiency and rarely lipoidol embolism to the brain and lung. The main utility of TACE is as the first line of treatment for inoperable HCC and as a bridge to liver transplant.

V. GENE THERAPY STRATEGIES AVAILABLE TO TARGET THE TUMOR AND THE TUMOR-ASSOCIATED VASCULATURE

While regional therapies, such as those discussed above, play an important role in managing metastatic disease, they are only applicable for patients with disease confined to a single region. Unfortunately, most patients with stage IV disease have evidence of metastatic spread in multiple locations. For these patients, it would be ideal to have a therapy that could be administered systemically, yet exert its activity regionally to mitigate systemic toxicity. Targeted gene therapy may allow for such a strategy to be realized.

   Over the last 20 years, since the first gene therapy studies were initiated, there have been more than 300 human gene therapy protocols approved or under review in the United States alone (Rosenberg *et al.*, 1990, 1999). Strategies to target the tumor and its vasculature using gene therapy have varied from simple methods of local intratumoral injection of gene therapy vectors to the development of novel genetically engineered viral and nonviral vectors to target the tumor following systemic administration. To date a number of cancer gene therapies have been designed and tested and these can be divided into five main categories: (1) suicide gene therapy, (2) rehabilitation of aberrant cell cycle, (3) immunomodulatory gene therapy, (4) antiangiogenesis gene therapy, and (5) oncolytic gene therapy.

   Angiogenesis, or the development of new vasculature, has been recognized as a critical step in the growth, invasion, and spread of tumors (Feldman *et al.*, 2000). A number of angiogenesis inhibitors have been utilized for gene therapy in a variety of models (see Table 7.1).

   One of the key regulators in the onset of tumor angiogenesis is the VEGF-A/VEGFR-2 axis. This has led to development of several agents targeting VEGF-A-mediated signal transduction, of which a few are in routine clinical use and over 20 are in clinical trials (Ellis and Hicklin, 2008).

Table 7.1. Antiangiogenic Agents in Clinical trials

Drug	Target	Published Clinical Trials
Bevacizumab (Avastin)	VEGF	Phase I, II, III
VEGF-Trap	VEGF	Recruiting
NM-3	VEGF	Recruiting
AE-941 (Neovastat)	VEGF, MMP	Phase I, II
IMC-ICII	VEGFR-2	Phase I
SU5416	VEGFR-2	Phase I, II
SU6668	VEGFR-2	Phase I
SU11248	VEGFR-1/2, PDGFR, KIT, FLT3	Phase I
PTK787/ZK222584	VEGFR-1/2	Phase I
ZD6474	VEGFR-2, EGFR	Recruiting
CP-547,632	VEGFR-2, FGFR-2, PDGFR	Recruiting
Endostatin	Various	Phase I
Angiostatin	Various	Phase I
TNP-470	Methionine aminopeptidase-2	Phase I
Thrombospondin-I (ABT-510)	CD36	Recruiting
Vitaxin	$\alpha v/\beta 3$	Phase I
EMD121974 (Cilengitide)	$\alpha v/\beta 3$, $\alpha v/\beta 5$	Phase I
Combretastatin A4	Endothelial tubulin	Phase I
ZD6126	Endothelial tubulin	Recruiting
2-Methoxyestradiol (2-ME)	microtubule	Recruiting
DMXAA	TNF-α induction	Phase I
Thalidomide	Various	Phase I, II, III
BMS-275291	MMP	Phase I
Celecoxib	COX-2	Phase I, II, III

A composite list of various antiangiogenic agents in clinical trials and their targets is shown. VEGF, vascular endothelial growth factor; VEGFR, VEGF receptor; MMP, matrix metalloproteinases; PDGFR, platelet derived growth factor receptor; EGFR, epidermal growth factor receptor; FGFR, fibroblasts growth factor receptor. Source Tandle *et al.* (2004)

The application of these agents as recombinant proteins for clinical tumor therapy has met with some logistical barriers. The main limiting factors have been (1) high-dose requirements, (2) manufacturing constraints, (3) relative instability of the recombinant proteins, and (4) limited antiangiogenic affect with the maximum tolerated systemic dose. Antiangiogenic gene therapy overcomes these limitations by utilizing host tissues as the site to produce higher concentrations of recombinant proteins locally for maximum antitumor activity. Characteristics of an ideal gene delivery vector would be tumor selectivity, minimal to zero systemic side effects, and effective gene transmission.

Vascular endothelial cells (ECs) differentially express surface receptors according to their physiologic location and functional state. The identification of the markers that define this diversity has led to the development of novel vector targets (Trepel *et al.*, 2008).

Safe and efficient DNA delivery systems are needed to overcome a range of extra- and intracellular transport barriers. Good delivery vectors demonstrate tissue specificity in delivering the targeted gene or drug to the site of action, that is, tumor and its vasculature. Minimizing nonspecific interactions, endosomal buffering, and DNA binding improves their efficiency in gene transduction and transfection.

A. Viral vectors

Viral vectors are effective gene delivery systems, as they are stable and produce higher quantities of the gene product compared to the systemic infusion of DNA or protein. There are five main classes of clinically applicable viral vectors; adenoviruses, vaccinia viruses (VVs), adeno-associated viruses (AAVs), retroviruses, lentiviruses, and herpes simplex-1 viruses (HSV-1s). Retroviruses and lentiviruses differ from the rest as they can integrate into the genome, infect only dividing cells, and have long-term transgene expression. The nonintegrated viral vectors can infect nondividing cells and exist predominantly as extrachromosomal episomes (Tandle et al., 2004).

1. Adenoviruses

Adenovirus is a double-stranded DNA virus. Most vectors have been made replication-defective to be suitable for cancer gene therapy and to decrease systemic toxicity. At present, there are 42 serotypes of adenovirus known to infect humans, out of which serotypes 2 and 5 are most often used for gene therapy (Liu and Deisseroth, 2006). The initial process of transduction involves binding of the head domain of the viral fiber protein to a specific cell-surface receptor called the Coxsackievirus-adenovirus receptor (CAR), followed by viral internalization and transport of the viral DNA into the cell nucleus through the virus penton base and cellular $\alpha\nu\beta3$ and $\alpha\nu\beta5$ integrin receptors (Bergelson et al., 1997; Wickham et al., 1993). Using the encapsulated adenovirus in bilamellar cationic liposomes can result in the transduction of CAR negative cells. These liposomes demonstrate resistance to the neutralizing antiadenoviral antibodies (Yotnda et al., 2002).

Adenovirus is a favored vector for gene therapy due to its effects on a wide variety of cell types and high gene transfer efficiency. They are genetically modified in multiple stages to enhance their safety as vectors for gene therapy. These modifications involve (1) incorporation of different tissue-specific promoters and enhancers to limit expression of the therapeutic gene, (2) removal of native receptors and the addition of cell- or tissue-specific ligands to target the cell or tissue of choice, and (3) utilization of suicide genes such as inducible caspase (iCaspase) to induce apoptosis in the infected cells (Song et al., 2005; Varda-Bloom et al., 2001).

Song *et al.* (2005) constructed AdhVEGFR2-iCaspase-9 adenoviral vector, capable of mediating expression of iCaspase-9 specifically in proliferating blood vessels and demonstrated its antiangiogenic effects *in vivo*.

The advantage of adenovirus as a vector for gene therapy includes the fact that they are efficient at *in vivo* gene transfer and produce robust levels of the transgene but do so transiently, especially in the liver after systemic injection. However, the main drawback of adenovirus is this transient expression. These short-term effects are explained by the nature of adenovirus as a nonintegrating vector. Expression is also limited by clearance of the immune system as well as receptor independent uptake by the reticuloendothelial system (RES).

2. Vaccinia virus

The reported ability of oncolytic VVs to infect tumor cells selectively and to express high levels of their transgene makes them an attractive vector for gene therapy (McCart *et al.*, 2004). One of the explanations for their tumor selectivity is their large molecular size (> 200 nm), selectively extravasating into the tumor interstitium favored by its leaky vasculature compared to normal, mature blood vessels.

These viruses have been used for their antitumor effects as well as for noninvasive monitoring of tumor and vector distribution. The creation of a double deleted thymidine kinase (TK) and vaccinia growth factor (VGF) deleted VV (vvDD–GFP) has improved tumor specificity and decreased systemic toxicity (McCart *et al.*, 2001). This mutant VV demonstrates a superior safety profile and a higher tumor selectivity compared to single deleted and wild type vectors.

Recombinant VV vectors expressing SSTR2 (human somatostatin receptor type 2) in combination with 111In radiolabeled, high-affinity synthetic peptide pentetreotide permits noninvasive molecular imaging of neuroendocrine tumors. This ability to track vectors noninvasively is important for monitoring vector distribution in gene therapy trials. Another major advantage of VV vectors is the ability to insert up to 25 kb of foreign DNA. This allows the vector to carry more than one gene, such as both reporter genes for imaging and therapeutic genes (McCart *et al.*, 2004).

3. Retroviruses

Retroviruses are linear single stranded RNA viruses with a lipid containing glycoprotein viral envelope. Upon entry into the cell, viral RNA is reverse-transcribed into DNA and randomly integrated into the host-cell genome (Miller, 1992). Their intrinsic selectivity for proliferating cells makes them suitable for targeting tumors. They can carry up to 8–9 kb of foreign DNA for transduction. Retroviruses are among the first viral vectors to be used in human cancer gene therapy. The most widely studied retroviral vector is the moloney murine leukemia virus (MoMuLV) (Seth, 2005).

Retroviruses have been genetically modified with several specific promoters such as Flt-1, ICAM-2, and KDR to target tumor-associated ECs. Liu et al. constructed a retroviral vector targeting tumor vasculature by the addition of the ligand protein Asn-Gly-Arg (NGR) sequence. Such modification led to improved binding efficiency and transduction of the vector to both human umbilical vein ECs and KSY1 ECs (Liu et al., 2000). A recombinant retroviral vector expressing the HSV-TK gene driven by a hybrid endothelial specific PPE-1 long terminal repeat (LTR) combined with chemotherapeutic agents resulted in widespread tumor cell apoptosis and vascular disruption in xenograft tumor models (Mavira et al., 2005). The main disadvantage of retroviruses as a gene therapy vector for patients has been their role in causing secondary cancers such as leukemia, most likely due to integration mutagenesis (Li et al., 2002).

4. Lentiviruses

The lentivirus is a subclass of retrovirus. As such, infection is associated with permanent integration of the gene, and thus can affect nondividing cells and terminally differentiated cells such as neurons, macrophages, and hematopoietic stem cells (Barker and Planelles, 2003). They have broad tissue tropism especially for transducing cells that lack receptors for adenovirus. So far, lentiviral vectors expressing matrix metalloproteinase-2 (MMP-2), angiostatin, and endostatin have been developed to target tumor vasculature (Shichinohe et al., 2001). Lentiviruses are also used to delivery therapeutic drugs to the tumor microenvironment such as, Interferon (IFN) under the control of the Tie-2 promoter to transduce hematopoietic stem/progenitor cells (HSPC). IFN in appropriate doses is known to exert antiproliferative, angiostatic, immune cell-activating functions, and significant inhibition of tumor growth (De Palma et al., 2008). However, lentiviruses have low transduction efficiency for ECs and may result in significant vector-associated cytotoxicity.

5. Herpes simplex-1 viruses

HSV is an enveloped, double strand DNA virus. HSV exhibits human cellular tropism by interacting with heparan sulfate molecules or nectin-1 through its viral glycoproteins gB and gC. Internalization of the virus occurs through the interaction of viral glycoprotein gD and the cellular FGF (fibroblast growth factor) receptor (Kaner et al., 1990).

IL-12 expressing HSV-1 vectors demonstrated antiangiogenic effects both in vivo and in vitro. The main advantage of HSV-1 as a gene therapy vector is its effect on a wide range of hosts and capacity to carry large genes for transgene insertion. Its toxicity has limited its use. The generation of inflammatory responses at the site of initial injection and distantly through neuronal transmission has limited its application for human gene therapy (Wood *et al.*, 1994).

6. Adeno-associated virus

The AAV is a single stranded DNA virus with an inherent defect in its ability to replicate. AAV remains in a latent state by integrating into the host-cell chromosomal DNA. AAV requires a helper virus such as adenovirus to replicate episomally and to initiate viral protein synthesis. At least eight serotypes of AAV are known, each with a distinct tissue tropism. Among the eight serotypes, AAV-2 has been studied extensively for its tropism for cell-surface heparan sulfate proteoglycan (HSPG) (Kashiwakura *et al.*, 2005). Other proposed coreceptors for AAV-2 are the $\alpha\nu\beta5$-integrin receptor, hepatocyte growth factor receptor c-Met, and the FGF receptor 1 (FGFR-1). AAV-5 serotype coreceptors are the platelet derived growth factor receptors (PDGFR-α and PDGFR-β). AAV has been used as a vector for the targeted delivery of angiostatin, mouse endostatin, and an antisense RNA against VEGF. The transduction of tumor cells such as glioblastoma (U87), hepatocellular carcinoma (Hep3B), and colorectal carcinoma (CX-1) with these recombinant AAV vectors demonstrated an ability to inhibit capillary EC growth and tumor cell VEGF secretion (Nguyen *et al.*, 1998).

AAV as a viral vector generates either a low or undetectable immune response from the host. Another advantage of AAV is that it can infect nondividing cells and has the potential for prolonged periods of transgene expression. Its limitations include its small transgene size capacity (5 kb) and its relative poor tropism for ECs resulting in poor transduction efficiency to tumor-associated ECs. Another explanation for the poor transduction of AAV is the sequestration of AAV-2 within the extracellular matrix and degradation of internalized AAV-2 particles by proteasomes (Nicklin *et al.*, 2001). The transduction efficiency of AAV vectors for ECs can be improved by the genetically engineered incorporation of EC targeted peptides expressed on their surface discovered by phage display techniques. Nicklin *et al.* (2001) engineered AAV-2 vectors by incorporating a heptamer peptide SIGYPLP into position I-587 of the AAV-2 capsid resulting in the efficient uptake of the vector by vascular ECs.

7. Adeno-associated viruses phage

Naturally occurring mammalian viral vectors are efficient gene transfer vectors but have a disadvantage of native tropism—which could be a drawback for targeted therapy and can lead to increased clearance by RES and immune stimulation. Over the past decade, research has concentrated on the development of targeted tissue-specific viral vectors created by genetic modification. This process involves the alteration of the viruses native tropism either by ablation or redirection to alternative receptors or both. Another method to generate tissue specificity has been the incorporation of homing peptides from bacteriophage display library screenings into mammalian viral vectors. However, this method resulted in an altered viral capsid structure and display within a viable viral capsid, affects the characteristics of targeting ligand peptides, and even prevent their display on the viral capsid altogether.

These limitations led to the creation of a novel hybrid vector that has the desirable properties of both mammalian viruses and bacterio-lytic viruses (bacteriophages). Bacteriophage display tissue-specific ligands but lack natural tropism for eukaryotic cells and are poor vectors for gene transfer.

Considering the above, Hajitou and colleagues have developed a hybrid prototype vector termed Adeno-Associated Virus Phage (AAVP), a hybrid of AAV and bacteriophage. They genetically incorporated compatible single stranded DNA of AAV2 cis-elements (such as ITR-inverted terminal repeats) into the phage DNA. Their targeted AAVP vector incorporated a mammalian transgene cassette flanked by full length ITRs of AAV-2 on RGD-4C peptide display phage that targets $\alpha\nu$ integrins (Hajitou et al., 2007).

TNF-α is an inflammatory cytokine with both antivascular and antitumor effects. TNF can also induce tissue factor production, increase vascular permeability, and increase procoagulant activity on ECs, resulting in severe systemic toxicity. At the maximum tolerated doses by systemic administration, there is no meaningful antitumor activity (Hagen and Eggermont, 2003). Techniques such as isolated organ perfusion limit the TNFα activity to the perfused tissue and have demonstrated antitumor activity, as exemplified by its use in the treatment of human primary melanomas, sarcomas, and metastatic hepatic tumors (Alexander et al., 1998; Grunhagen et al., 2005, 2006).

As an alternative to isolation perfusion techniques, a novel approach was developed by Tandle et al., utilizing AAVP viral vectors to deliver TNFα to the target tissue of interest through systemic administration. They reported significant tumor growth inhibition compared to their controls using a model of human melanoma. They also demonstrated significant activity of TNFα in the tumor tissue compared to no activity in the normal tissues such as liver and spleen. This approach retained the antitumor activity of TNF in the transduced tumor cells with minimal systemic toxicity (Tandle et al., 2009).

The role and pharmacodynamics of AAVP viral vectors in the targeted delivery of TNFα to tumor tissues were also demonstrated and assessed in the naturally occurring cancers of pet dogs. Dogs are an ideal animal model for studying the efficacy and toxicity of TNF as dogs demonstrate the same sensitivity to TNF as humans. The spontaneously occurring tumors in these animals are also a better model for tumors seen in human patients. This study demonstrated selective targeting of tumor-associated vasculature and sparing of normal tissues assessed via serial biopsy of both tumor and normal tissues. A variety of different tumor types were evaluated. At the optimal dose, repetitive dosing was done in a cohort of 14 dogs with unresectable tumors. Objective tumor regression was seen in two dogs (14%), stable disease in six (43%), and disease progression in six (43%). Responses were reported using the Response Evaluation Criteria in Solid Tumors (RECIST) criteria. This study provides a basic design that can be directly adapted for human phase I and phase II clinical trials (Paoloni et al., 2009).

Other phage displaying the cyclic peptide His-Try-Gly-Phe (HWGF) was created specifically to target angiogenic blood vessel in vivo, which inhibits MMP-2 and MMP-9 metalloproteinases (Tandle et al., 2004). MMP-2 has a defined role in tumor cell proliferation and angiogenesis by directly binding to $\alpha v \beta 3$ integrins expressed on proliferating ECs. Inhibition of such peptides utilizing these targeted display phage results in inhibition of tumor angiogenesis.

AAVP vectors form an attractive alternative to animal viral vectors, as they lack intrinsic tropism for mammalian cells and have improved mammalian transduction efficiency and effective gene transmission. This new generation of prokaryotic–eukaryotic vector can also be produced in large titers (Hajitou et al., 2007).

B. Nonviral particle-based delivery systems

While viral vectors offer significant promise, there may be advantages to nonviral vector systems. Such an approach may confer a lower risk of toxicity and immune response compared to viral vectors. An efficient and safe DNA delivery system needs to be developed to surmount both intra- and extracellular transport barriers as well as improve the levels of transgene expression.

1. Liposomes

Liposomes are structures composed of phospholipids and cholesterol. They have been widely used for the delivery of chemotherapeutic drugs such as adriamycin as well as plasmids, imaging agents, antigens, lipids, and DNA. Conjugation of liposomes with the long circulating polyethyleneglycol (PEG) has been used as a carrier for a variety of drugs. Attachment of specific antibodies or antibody fragments on their surface allows liposomes to target drugs to their site of action.

Liposomes can be targeted to a specific cell or tissue by conjugating with a protein interacting with a tissue-specific marker such as scFv for human ED-B. Rodolfo et al. (2001) showed that liposomal delivered angiostatin could inhibit the growth of melanoma tumors in mice. Liposomes modified with angiogenic homing peptides, such as PIVO-8 and PIVO-24 targeting peptides for ECs, can also suppress tumor growth (Chang *et al.*, 2009). The preferential uptake of liposomes could be used for tumor diagnostic imagining and also therapeutic delivery of cyototoxic drugs.

Liposomes are more stable and have a higher rate of gene transfer efficiency than naked DNA molecules. Cationic liposome:DNA complexes have a 13–15-fold higher chance of transfection than the corresponding naked DNA (Chen *et al.*, 1999). Inherent properties of liposomes such as poor stability and low encapsulation efficiency limit their role as an ideal carrier for drugs or genes to target tissues.

2. Nanoparticles

Nanoparticles (NPs) are submicron sized polymeric colloidal particles with a therapeutic agent of interest encapsulated within their polymeric matrix or adsorbed or conjugated onto their surface. The process of gold NP synthesis was initially described by Faraday (1857). Gold NPs are inert molecules, which can carry proteins without changing their biological properties (Chandler *et al.*, 2001). The versatile nature of these NPs demonstrates many advantages ranging from DNA diagnostics and biosensors to the treatment of liver cancer and sarcoma (Mirkin *et al.*, 1996; Root *et al.*, 1954; Rubin and Levitt, 1964). There are three main components of a NP targeted delivery system, the core material, the active agent, and modifiers to improve efficacy. NPs invade into tumors either passively or by active targeting mechanisms. Passive accumulation of the NPs is facilitated by the leaky and disorganized vasculature of the tumor. In comparison, active targeting is based on the NPs utilizing a targeting moiety conjugated to the NP itself to allow for sequestration in the tumor.

TNF is a highly potent and toxic antitumor protein. Its systemic toxicity can be circumvented by targeting its delivery to the site of its action such as the tumor bed. This can be accomplished either by isolated organ perfusion or by targeted systemic approaches using delivery systems like gene vectors, biodegradable polymers, or colloidal gold NPs. Colloidal gold NPs are a good platform for the delivery of TNF protein to solid tumors. Initial formulations were bound directly to the NPs, which are less toxic compared to native TNF; however, efficacy was limited by rapid uptake and clearance in the RES (Furman *et al.*, 1993; Schiller *et al.*, 1991). When the particles were reformulated with thiol-derivatized PEG (PEG-THIOL) bound to the TNF molecule on the NPs surface, it avoided sequestration by the RES and subsequently resulted in the effective delivery of TNF protein to the solid tumor. A number of TNF molecules are bound to a single colloidal gold NP, which further facilitates TNF sequestration (Paciotti *et al.*, 2004).

CYT-6091 is a new formulation of pegylated colloidal gold-TNF that evades the RES, thus improving the efficiency of TNF molecule delivery. Efficacy and toxicity studies were carried out in preclinical trials using colon cancer xenograft models. These studies demonstrated tumor regression in 72% and 82% of animals without any adverse toxic effects. After demonstrating the efficacy and safety of the CYT-6091 construct in preclinical studies, a phase I clinical trial was performed in patients with cancer refractory to standard therapies. Trial eligibility was limited to advanced-stage solid malignancies and conducted as a dose-escalation trial. The primary end point of the study was to determine the maximum tolerated dose and secondary end points were to assess disease response, pharmacokinetic data, and assess for the presence of gold particles in the tumor and adjacent normal tissue (obtained through biopsy). A preliminary report demonstrated no dose-limiting toxicity after administration of CYT-6091 and biopsy analysis showed 10-fold increase in the number of gold NPs in tumors compared to adjacent normal tissue. The data from the full trial have not yet been reported (Powell et al., 2010).

The two main challenges faced by the particle-based delivery systems are escaping the RES system and the high interstitial fluid pressures found in solid tumors (Netti et al., 1999; Rofstad, 2002; Stohrer et al., 2000; Zhang et al., 2000). Other considerations are their nonspecific uptake, poor adsorption, short half-life, and low in vivo potency for cell transfection (Tandle et al., 2004).

VI. CONCLUSIONS

There are a variety of strategies available to target the tumor and its associated vasculature. The benefits of this approach are to maximize antitumor activity while minimizing systemic toxicity. Further research and development is needed to develop the ideal vector, which remains the backbone for targeted therapy. Novel vectors and particles, engineered to be delivered systemically and to traffic selectively, are the future of cancer therapy.

References

Alexander, H. R., Jr, Bartlett, D. L., Libutti, S. K., Fraker, D. L., Moser, T., and Rosenberg, S. A. (1998). Isolated hepatic perfusion with tumor necrosis factor and melphalan for unresectable cancers confined to the liver. *J. Surg. Oncol.* **16,** 1479–1489.

Alexander, H. R., Jr., Libutti, S. K., Pingpank, J. F., et al. (2005). Isolated hepatic perfusion for the treatment of patients with colorectal cancer liver metastases after irinotecan-based therapy. *Ann. Surg. Oncol.* **12,** 138–144.

Barker, E., and Planelles, V. (2003). Vectors derived from the human immunodeficiency virus, HIV-1. *Front. Biosci.* **8,** 491–510.

Benckhuijsen, C., Kroon, B. B., van Geel, A. N., and Wieberdink, J. (1988). Regional perfusion treatment with melphalan for melanoma in a limb: An evaluation of drug kinetics. *Eur. J. Surg. Oncol.* **14,** 157–163.

Bergelson, J. M., Cunningham, J. A., Droguett, G., *et al.* (1997). Isolation of a common receptor for coxsackie B viruses and adenoviruses 2 and 5. *Science* **275,** 1320–1323.

Chandler, J., Robinson, N., and Whiting, K. (2001). Handling false signals in gold-based tests. *I.V.D Technol.* **72,** 34–35.

Chang, D., Chiu, C., Kuo, S., Lin, W., Lo, A., Wang, Y., Li, Pi., and Wu, H. (2009). Antiangiogenic targeting liposomes increase therapeutic efficacy for solid tumors. *J. Biol. Chem.* **284**(19), 12905–12916.

Chen, Q. R., Kumar, D., Stass, S. A., and Mixson, A. J. (1999). Liposomes complexed to plasmids encoding angiostatin and endostatin inhibit breast cancer in nude mice. *Cancer Res.* **59,** 3308–3312.

Creech, O., Jr., Krementz, E. T., Ryan, R. F., and Winblad, J. N. (1958). Chemotherapy of cancer: Regional perfusion utilizing an extracorporeal circuit. *Ann. Surg.* **148,** 616–632.

De Palma, M., Mazzieri, R., Politi, L. S., Pucci, F., Zonari, E., Sitia, G., *et al.* (2008). Tumor targeted interferon-alpha delivery by Tie2-expressing monocytes inhibits tumor growth and metastasis. *Cancer Cell* **14,** 299–311.

Eggermont, A. M., Schraffordt Koops, H., Klausner, J. M., Kroon, B. B., Schlag, P. M., Lienard, D., van Geel, A. N., Hoekstra, H. J., Meller, I., Nieweg, O. E., Kettelhack, C., *et al.* (1996a). Isolated limb perfusion with tumor necrosis factor and melphalan for limb salvage in 186 patients with locally advanced soft tissue extremity sarcomas. The cumulative multicenter European experience. *Ann. Surg.* **224,** 756–764.

Eggermont, A. M., Schraffordt Koops, H., Lienard, D., Kroon, B. B., van Geel, A. N., Hoekstra, H. J., and Lejeune, F. J. (1996b). Isolated limb perfusion with high- dose tumor necrosis factor-alpha in combination with interferon-gamma and melphalan for nonre- sectable extremity soft tissue sarcomas: A multi-center trial. *J. Clin. Oncol.* **14,** 2653–2665.

Eggermont, A. M., Schraffordt Koops, H., Klausner, J. M., Schlag, P. M., Lienard, D., Kroon, B. B. R., Gustafson, P., Steinmann, G., Clarke, J., and Lejeune, F. (1999). Limb salvage by isolated limb perfusion (ILP) with TNF and melphalan in patients with locally advanced soft tissue sarcomas: Outcome of 270 ILPs in 246 patients. *Proc. Am. Soc. Clin. Oncol.* **18,** 2067.

Ellis, L. M., and Hicklin, D. J. (2008). VEGF-targeted therapy: Mechanisms of anti-tumour activity. *Nat. Rev. Cancer* **8,** 579–591.

Faraday, M. (1857). Experimental relations of gold (and other metals) to light. *Philos. Trans. R. Soc. London* **14,** 145–181.

Feldman, A. L., Restifo, N. P., Alexander, H. R., Bartlett, D. L., Hwu, P., Seth, P., and Libutti, S. K. (2000). Antiangiogenic gene therapy of cancer utilizing a recombinant adenovirus to elevate systemic endostatin levels in mice. *Cancer Res.* **60,** 1503–1506.

Furman, W. I., Strother, D., McClain, K., Bell, B., Leventhol, B., and Pratt, C. B. (1993). Phase I clinical trial of recombinant human tumor necrosis factor in children with refractory solid tumors: A pediatric oncology group study. *J. Clin. Oncol.* **11,** 2205–2210.

Grunhagen, D. J., Brunstein, F., Graveland, W. J., van Geel, A. N., de Wilt, J. H., and Eggermont, A. M. (2005). Isolated limb perfusion with tumor necrosis factor and melphalan prevents amputation in patients with multiple sarcomas in arm or leg. *Ann. Surg. Oncol.* **12,** 473–479.

Grünhagen, D. J., de Wilt, J. H., Graveland, W. J., Verhoef, C., van Geel, A. N., and Eggermont, A. M. (2006). Outcome and prognostic factor analysis of 217 consecutive isolated limb perfusions with tumor necrosis factor-alpha and melphalan for limb-threatening soft tissue sarcoma. *Cancer* **106,** 1776–1784.

Gutman, M., Inbar, M., Lev-Shlush, D., Abu-Abid, S., Mozes, M., Chaitchik, S., Meller, I., and Klausner, J. M. (1997). High dose tumor necrosis factor-alpha and melphalan administered via isolated limb perfusion for advanced limb soft tissue sarcoma results in a >90% response rate and limb preservation. *Cancer* **79**, 1129–1137.

Hagen, T. L., and Eggermont, A. M. (2003). Solid tumor therapy: Manipulation of the vasculature with TNF. *Technol. Cancer Res. Treat.* **2**, 195–203.

Hajitou, A., Range, L. R., Trepel, M., *et al.* (2007). Design and construction of targeted AAVP vectors for mammalian cell transduction. *Nat. Protoc.* **2**, 523–531.

Hill, S., Fawcett, W. J., Sheldon, J., Soni, N., Williams, T., and Thomas, J. M. (1993). Low-dose tumour necrosis factor alpha and melphalan in hyperthermic isolated limb perfusion. *Br. J. Surg.* **80**, 995–997.

Kaner, R. J., Baird, A., Manusukhani, A., *et al.* (1990). Fibroblast growth factor receptor is a portal of cellular entry for herpes simplex virus type-1. *Science* **248**, 1410–1413.

Kashiwakura, Y., Tamayose, K., Iwabuchi, K., *et al.* (2005). Hepatocyte growth factor receptor is a coreceptor for adeno-associated virus type 2 infection. *J. Virol.* **79**, 609–614.

Li, Z., Düllmann, J., Schiedlmeier, B., Schmidt, M., Stocking, C., Wahlers, A., Ostertag, W., Kühlcke, K., Fehse, B., Baum, C., *et al.* (2002). Murine Leukemia Induced by Retroviral Gene Marking. *Science* **296**, 497.

Libutti, S. K., Bartlett, D. L., Fraker, D. L., *et al.* (2000). Technique and results of hyperthermic isolated hepatic perfusion with tumor necrosis factor and melphalan for the treatment of unresectable hepatic malignancies. *J. Am. Coll. Surg.* **191**, 519–530.

Liu, Y., and Deisseroth, A. (2006). Oncolytic adenoviral vector carrying the cytosine deaminase gene for melanoma gene. *Cancer Gene Therapy* **13**, 845–855.

Liu, L., Anderson, W. F., Beart, R. W., Gordon, E. M., and Hall, F. L. (2000). Incorporation of tumor vasculature targeting motifs into moloney murine leukemia virus env escort proteins enhances retrovirus binding and transduction of human endothelial cells. *J. Virol.* **74**, 5320–5328.

Mavira, G., Harrington, K. J., Marshall, C. J., and Porter, C. D. (2005). In vivo efficacy of HSV-TK transcriptionally targeted to the tumor vasculature is augmented by combination with cytotoxic chemotherapy. *J. Gene Med.* **7**, 263–275.

McCart, J. A., Ward, J. M., Lee, J., Hu, Y., Alexander, H. R., Libutti, S. K., Moss, B., and Bartlett, D. L. (2001). Systemic cancer therapy with a tumor-selective vaccinia virus mutant lacking thymidine kinase and vaccinia growth factor genes. *Cancer Res.* **61**(24), 8751–8757.

McCart, J. A., Mehta, N., Scollard, D., Reilly, R. M., Carrasquillo, J. A., Tang, N., Deng, H., Miller, M., Xu, H., Libutti, S. K., Alexander, H. R., and Bartlett, D. L. (2004). Oncolytic vaccinia virus expressing the human somatostatin receptor SSTR2: Molecular imaging after systemic delivery using 111In-pentetreotide. *Mol. Ther.* **10**(3), 553–561.

Miao, N., Pingpank, J. K., Alexander, R. H., Steinberg, S. M., Beresneva, T., and Quezado, Z. (2008). Percutaneous hepatic perfusion in patients with metastatic liver cancer: Anesthetic, hemodynamic, and metabolic considerations. *Ann. Surg. Oncol.* **15**(3), 815–823.

Miller, A. D. (1992). Retroviral vectors. *Curr. Top. Microbiol. Immunol.* **158**, 1–24.

Mirkin, C. A., Letsinger, R. L., Mucic, R. C., and Storhoff, J. J. (1996). A DNA based method for rationally assembling nanoparticles into macroscopic materials. *Nature* **382**, 607–609.

National Center for Health Statistics and Centers for Disease Control and Prevention (2006). US Mortality Public Use Data Tape 2003.

Netti, P. A., Hamberg, L. M., Babich, J. W., Kierstia, D., Graham, W., Hunter, G. J., Wolf, G. L., Fischnan, A., Boucher, Y., and Jain, R. K. (1999). Enhancement of fluid filtration across tumor vessels: Implication for delivery of macromolecules. *Proc. Natl. Acad. Sci. USA* **96**, 3137–3142.

Nguyen, J. T., Wu, P., Clouse, M. E., Hlatky, L., and Terwilliger, E. F. (1998). Adeno-associated virus-mediated delivery of antiangiogenic factors as an antitumor strategy. *Cancer Res.* **8**, 5673–5677.

Nicklin, S. A., Buening, H., Dishart, K., *et al.* (2001). Efficient and selective AAV2-mediated gene transfer directed to human vascular endothelial cells. *Mol. Ther.* **4**, 174–181.

Okayasu, I., Hatakeyama, S., Yoshida, T., Yoshimatsu, S., Tsuruta, K., Miyamoto, H., *et al.* (1988). Selective and persistent deposition and gradual drainage of iodized oil, Lipiodol in the hepatocellular carcinoma after injection into the feeding hepatic artery. *Am. J. Clin. Pathol.* **90**, 536–544.

Paciotti, G. F., Myer, L., Weinreich, D., Goia, D., Pavel, N., McLaughlin, R. E., and Tamarkin, L. (2004). Colloidal gold: A novel nanoparticle vector for tumor directed drug delivery. *Drug Deliv.* **11**(3), 169–183.

Paoloni, M. C., Tandle, A., Mazcko, C., Arap, W., Khanna, C., Pasqualini, R., Libutti, S. K., *et al.* (2009). Launching a novel preclinical infrastructure: Comparative oncology trials consortium directed therapeutic targeting ofTNFα to cancer vasculature. *PLoS One* **4**(3), e4972.

Pingpank, J. F., Libutti, S. K., Chang, R., Wood, B. J., Neeman, Ziv, Kam, Anthony, W., Figg, William, D., Geoffrey, D., Alexander, H. R., *et al.* (2005). Phase I study of hepatic arterial melphalan infusion and hepatic venous hemofiltration using percutaneously placed catheters in patients with unresectable hepatic malignancies. *J. Clin. Oncol.* **23**, 3465–3474.

Powell, A. C., Paciotti, G. F., and Libutti, S. K. (2010). Colloidal gold: A novel nanoparticles for targeted cancer therapeutics. *Methods Mol. Boil.* **624**, 375–384.

Rodolfo, M., Cato, E. M., Soldati, S., Ceruti, R., Asioli, M., Scanziani, E., Vezzoni, P., Parmiani, G., and Sacco, M. G. (2001). Growth of human melanoma xenografts is suppressed by systemic angiostatin gene therapy. *Cancer Gene Ther.* **8**, 491–496.

Rofstad, E. K., Siv, H., Tunheim, Berit, M., Bjørn, A., Graff, E. F., Halsør, Kristin, N., and Galappathi, K. (2002). Pulmonary and lymph node metastasis is associated with primary tumor interstitial fluid pressure in human melanoma xenografts. *Cancer Res.* **62**, 661–664.

Root, S. W., Andrews, G. A., Knieseley, R. M., and Tyor, M. P. (1954). The distribution and radiation effects of intravenously administered colloidal Au198 in man. *Cancer* **7**, 856–866.

Rosenberg, S. A., Aebersold, P., Cornetta, K., Kasid, A., Morgan, R. A., Moen, R., *et al.* (1990). Gene transfer into humans immunotherapy of patients with advanced melanoma, using tumor-infiltrating lymphocytes modified by retroviral gene transduction. *N. Engl. J. Med.* **323**, 570–578.

Rosenberg, S. A., Blaese, R. M., Brenner, M. K., Deisseroth, A. B., Led- ley, F. D., Lotze, M. T., *et al.* (1999). Human gene marker/therapy clinical protocols. *Hum. Gene. Ther.* **10**, 3067–3123.

Rubin, P., and Levitt, S. H. (1964). The response of disseminated reticulum cell sarcoma to the intravenous injection of colloidal radioactive gold. *J. Nuclear Med.* **5**, 581–594.

Schiller, J. H., Storer, B. E., Witt, P. L., Alberti, D., Tombes, M. M. B., Arzoomanian, R., Proctor, R. A., McCarthy, D., Brown, R. R., Voss, S. D., Remick, S. C., Grem, J. L., Borden, E. C., and Trump, D. L. (1991). Biological and clinical effects of intravenous tumor necrosis factor-a administered three times weekly. *Cancer Res.* **51**, 1651–1658.

Seth, P. (2005). Vector-medicated cancer gene therapy. *Cancer Biol. Ther.* **4**(5), 512–517.

Shichinohe, T., Bochner, B. H., Mizutani, K., Nishida, M., Hegerich-Gilliam, S., Naldini, L., and Kasahara, N. (2001). Development of lentiviral vectors for antiangiogenic gene delivery. *Cancer Gene Ther.* **8**, 879–889.

Song, W., Sun, Q., Dong, Z., Spencer, D. M., Nunez, G., and Nor, J. E. (2005). Antiangiogenic gene therapy: Disruption of neovascular networks mediated by inducible caspase-9 delivered with a transcriptionally targeted adenoviral vector. *Gene Ther.* **12**, 320–329.

Stohrer, M., Boucher, Y., Stangassinger, M., and Jain, R. K. (2000). Oncotic pressure in solid tumors is elevated. *Cancer Res.* **60**, 4251–4255.

Tandle, A., Blazer, D. G., III, and Libutti, S. K. (2004). Antiangiogenic gene therapy of cancer: Recent developments. *J. Transl. Med.* **2**, 22.

Tandle, A., Hanna, E., Lorang, D., Hajitou, A., Moya, C., Pasqualini, R., Arap, W., Adem, A., Straker, E., Hewitt, S., and Libutti, S. K. (2009). Tumor vasculature-targeted delivery of tumor necrosis factor-α. *Cancer* **115**(1), 128–139.

Trepel, M., Pasqualini, E., and Arap, W. (2008). Screening phage-display peptide libraries for vascular-targeted peptides. *Methods Enzymol.* **445,** 88–106.

Varda-Bloom, N., Shaish, A., Gonen, A., *et al.* (2001). Tissue- specific gene therapy directed to tumor angiogenesis. *Gene Ther.* **8,** 819–827.

Wickham, T. J., Mathias, P., Cheresh, D. A., and Nemerow, G. R. (1993). Integrins alpha v beta 3 and alpha v beta 5 promote adenovirus internalization but not virus attachment. *Cell* **73,** 309–319.

Wieberdink, J., Benckhuysen, C., Braat, R. P., van Slooten, E. A., and Olthuis, G. A. (1982). Dosimetry in isolation perfusion of the limbs by assessment of perfused tissue volume and grading of toxic tissue reactions. *Eur. J. Cancer Clin. Oncol.* **18,** 905–910.

Wigmore, S. J., Redhead, D. N., Thomson, B. N., Currie, E. J., Parks, R. W., Madhavan, K. K., *et al.* (2003). Postchemoembolisation syndrome—tumour necrosis or hepatocyte injury? *Br. J. Cancer* **89,** 1423–1427.

Wood, R. J., Byrnes, A. P., Pfaff, D. W., Rabkin, S. D., and Charlton, H. M. (1994). Inflammatory effects of gene transfer into the CNS with defective HSV-1 vectors. *Gene Ther.* **1,** 283–291.

Yotnda, P., Chen, D. H., Chiu, W., Piedra, P. A., Davis, A., Templeton, N. S., and Brenner, M. K. (2002). Bilamellar cationic liposomes protect adenovectors from preexisting humoral immune responses. *Mol. Ther.* **5,** 233–241.

Zhang, X. Y., Luck, J., Derhirst, M. W., and Fan Yuan, F. (2000). Interstitial hydraulic conductivity in a fibrosarcoma. *Am. J. Physiol. Heart Circ. Physiol.* **279,** H2726–H2734.

Index

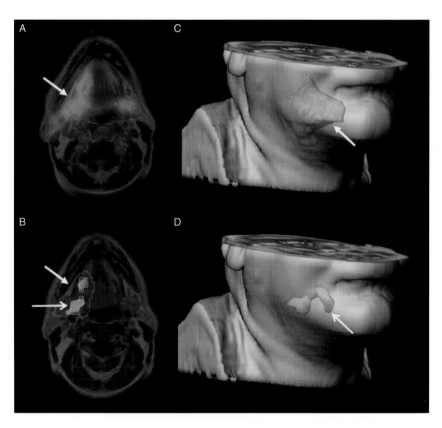

Chapter 1, Figure 1.5 (See Page 21 of this volume).

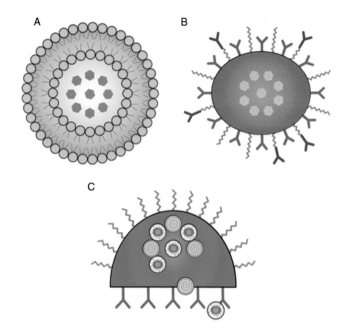

Chapter 2, Figure 2.1 (See Page 34 of this volume).

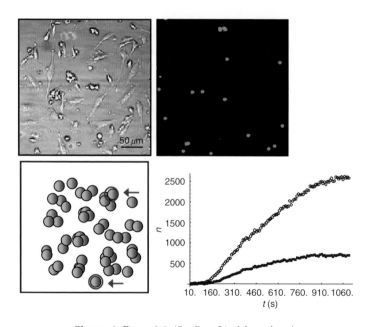

Chapter 2, Figure 2.9 (See Page 54 of this volume).

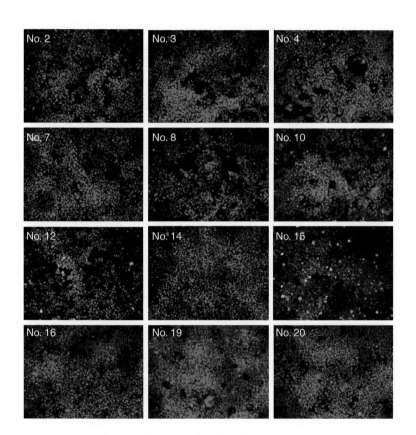

Chapter 4, Figure 4.1 (See Page 88 of this volume).

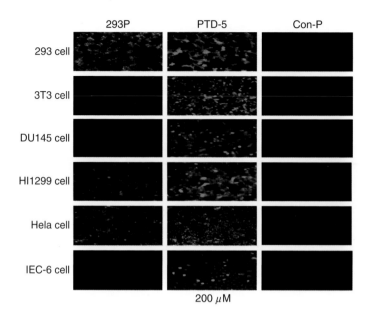

Chapter 4, Figure 4.2 (See Page 89 of this volume).

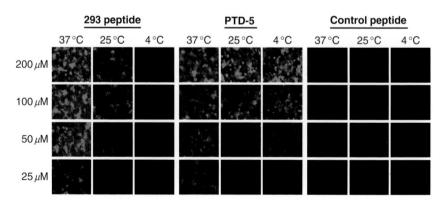

Chapter 4, Figure 4.3 (See Page 89 of this volume).